Chemical Thermodynamics in Materials Science

Taishi Matsushita · Kusuhiro Mukai

Chemical Thermodynamics in Materials Science

From Basics to Practical Applications

Taishi Matsushita
Department of Materials and Manufacturing, School of Engineering
Jönköping University
Jönköping
Sweden

Kusuhiro Mukai
Kyushu Institute of Technology
Kitakyushu
Japan

ISBN 978-981-13-4407-7 ISBN 978-981-13-0405-7 (eBook)
https://doi.org/10.1007/978-981-13-0405-7

Printed on acid-free paper

This Springer imprint is published by the registered company Springer Nature Singapore Pte Ltd. part of Springer Nature
The registered company address is: 152 Beach Road, #21-01/04 Gateway East, Singapore 189721, Singapore

Preface

This book, *Chemical Thermodynamics in Materials Science*, was originally developed as course material at the School of Engineering, Jönköping University, Sweden, based on a book written by one of the authors (KM) in 1992 (the English title of which is 'How to Utilize Chemical Thermodynamics'; Kyoritsu Shuppan Co., Ltd., Tokyo).

Although the Japanese book was widely accepted by many students and researchers, a quarter of a century has passed, and in the intervening time, the author has received many comments and much input from readers. This volume retains the contents of its predecessor, but many flaws have been corrected and new chapters added. Specifically, the chapter on entropy has been significantly revised, and the sections on the Carnot cycle and basic concepts of statistical thermodynamics have been added. In addition, thermodynamic calculation software packages such as Thermo-Calc have become more widely used in recent years, and so a chapter on the basics of computational thermodynamics has also been added.

The above-mentioned Japanese book assumes that readers have a basic knowledge of thermodynamics (beginning with the practical applications of the van't Hoff isotherm), but the structure of this book has been changed extensively so that readers can learn thermodynamics step by step, from beginning to end. Here, many schematic diagrams and examples are presented to facilitate understanding of peculiar concepts in thermodynamics.

Through our teaching activities, we realised that one difficulty experienced by those beginning to study thermodynamics is the symbols used in books, as different texts on thermodynamics use different symbols for the same concepts, or the same symbol for different concepts. This creates confusion for beginners, and so in this book the symbols used are those recommended by the International Union of Pure and Applied Chemistry (IUPAC).

In order to improve metallurgical processes by applying knowledge of thermodynamics and assessing the calculation results of thermodynamic software packages, a systematic and correct understanding of thermodynamics is required. However, books from which one can learn thermodynamics from the basic to advanced levels are rarely published. This book bridges the gap between the basic elements of thermodynamics, which are covered in general books on the subject, and their applications, which are partially discussed within specialised texts written for a specific field.

This book is written with a focus on learning reliable, applied skills using easy-to-understand explanations and in-depth descriptions and schematic diagrams. It can be used to teach the basics of chemical thermodynamics and their applications to beginners, but can also be used by advanced readers (postgraduate students/researchers) to learn or refresh the basic concepts of the subject.

We hope that this book helps you to understand and utilise in practice the basic elements of chemical thermodynamics.

Jönköping, Sweden

Taishi Matsushita
Associate Professor, Jönköping University, Sweden

Fukuoka, Japan
May 2018

Kusuhiro Mukai
Professor Emeritus, Kyushu Institute of Technology, Japan

Contents

Chapter 1
Introduction

Today, chemical thermodynamics is as well-established an academic field as Newtonian mechanics and, as our interest has now turned to how to best utilise it, thermodynamic databases containing values ranging between room temperature and elevated temperatures have been developed. However, when attempting to utilise thermodynamics practically, we often find ourselves faced with concepts, difficulties, conventions, and rules the nature which may cause us to hesitate to put them to use or even lead us to use them improperly, despite the fact that they present us with relatively small mathematical difficulties.

Let us provide a brief introduction to thermodynamics. If the state of a system can be described only as a function of work, W, we can approach it using Newtonian mechanics; if it can be described only as a function of heat, Q, we can understand it in relation to thermochemistry. In a scenario in which both work and heat are required in order to describe the state of a system, however, we need thermodynamics, which allows us to ascertain the equilibrium position—a function of W, Q, and mass, m (Chap. 7)—of a system. If we know the equilibrium position, we know in which direction a reaction will proceed in real systems in specific states (Chap. 10). Thermodynamics is a reliable academic field, and as long as we recognise the scope to which thermodynamic concepts are applicable, the results are never wrong.

Several functions (variables) in chemical thermodynamics are required to find the equilibrium state of a system. These include the Gibbs energy, G, and Helmholtz energy, F. G is used to find the equilibrium position of a system when temperature, T, and pressure, P, are constant (Sect. 7.1.5). F is used to find the equilibrium position of a system when T and volume, V, are constant (Sect. 7.1.4). Hence, G can be used for a system which includes a gas phase, and F can be used for a system which consists of liquid and/or solid phases. Both G and F are termed 'energy', but their values decrease with increasing T, and the slope of energy against temperature is $-S$ (where S is the entropy; Sects. 7.1.4, 7.1.5). S is also a unique and interesting function, and can be quantitatively described as *a function to describe the degree of randomness of a system* (Sect. 6.1).

T. Matsushita and K. Mukai, *Chemical Thermodynamics in Materials Science*,
https://doi.org/10.1007/978-981-13-0405-7_1

The S of a system increases when a reaction (a system shift) occurs spontaneously at a constant energy (mechanical work and heat energy) and constant mass (Chap. 6). This can be easily understood through everyday examples that an environment is littered with, wherein systems tend towards becoming chaotic if left in a natural state; if the increase in S (or entropy term TS) is larger than the increase in energy of the system, the system shift occurs spontaneously, even if the energy of the system increases.

In order to understand the above-mentioned concepts in thermodynamics, we will start with the basics—the first (Chap. 3) and second (Chap. 5) laws of thermodynamics—before going into more detail.

What, then, can be achieved or understood through an understanding of chemical thermodynamics?

The two primary answers to this, from the point of view of a researcher or engineer, are as follows:

(1) We can know the direction and equilibrium position of a reaction.
(2) We can provide guidelines for discussing and understanding experimental results.

By ascertaining the equilibrium position of a reaction, we know the conditions that enhance or suppress it. It is impossible for a new substance (e.g. a compound) to be formed if the process intended to form it violates the laws of thermodynamics—and the reliability and usefulness of thermodynamics rests on this. However, thermodynamics does not tell us how fast a reaction occurs, and this is one of the limitations of the field and an aspect that must be considered in order to properly utilise it.

For example, experimental results must satisfy several conditions, including Hess's law (Sect. 4.7); if they do not, something is wrong with the calculation or experimental process. According to the Clausius-Clapeyron equation (Sect. 7.3.2), the relationship between log P (vapour pressure) and $1/T$ should be linear within a relatively small temperature range. For a phase diagram, it must not violate the phase rule (Sect. 7.5).

Generally speaking, it is important to recognise the scope of science. Chemical thermodynamics is no exception to this, and so we should understand the following points:

(1) Only equilibrium states can be discussed.
(2) Only macroscopic systems can be discussed.
(3) The equations/laws of thermodynamics can only be applied using certain assumptions/restrictions.

Regarding (1); it is common to expand the discussion to non-equilibrium states, in spite of the fact that chemical thermodynamics are only applicable to equilibrium states. The academic field of non-equilibrium thermodynamics, proposed by L. Onsager in the 1930s, has been greatly developed by several of Onsager's predecessors, but can still only be applied to some non-equilibrium systems.

Regarding (2); thermodynamics cannot be applied to a small system whose state is influenced by the behaviour of individual atoms and molecules.

Regarding (3); this is a statement that is generally true, but often forgotten in relation to thermodynamics.

One may hesitate to apply thermodynamics due to a lack of understanding of the specific concepts, and understanding the core of any given field is essential to utilising and making best use of it. Using this book, the reader will learn the above-mentioned concepts of thermodynamics through schematic diagrams, examples, and rigorously-derived equations.

Chapter 2
Symbols and Glossary

Different books on chemical thermodynamics use many similar symbols, and sometimes the same symbols are used for different things, creating confusion for beginners. In addition, there are many terms and concepts in thermodynamics that seem to be similar, leading to further uncertainty. In this chapter, the symbols and terms used—-particularly those that might create difficulties for beginners—are summarised.

2.1 Symbols

The major symbols used in this book are listed here. They are based on the 2008 IUPAC recommendations (Cohen et al. 2008)—the so-called 'Green Book'— which is recognised as an international standard.

Latin alphabet

a_i :	Activity of component i
$a_i^{(\mathrm{H})}$:	Activity of component i(Henrian (1 mass% i) standard state)
$a_i^{(\mathrm{H})\prime}$:	Activity of component i(Henrian standard state)
$a_i^{(\mathrm{R})}$:	Activity of component i (Raoultian standard state)
$\mathcal{C}$:	Number of components (chemical species) in the system
c_P :	Specific heat capacity at constant pressure
C_P :	Heat capacity at constant pressure
$\overline{C}_P$:	Molar heat capacity at constant pressure
c_V :	Specific heat capacity at constant volume
C_V :	Heat capacity at constant volume
$\overline{C}_V$:	Molar heat capacity at constant volume
d (in differentiation):	Derivative (infinitesimal change)
$\mathrm{e}_i^{(j)}$:	Interaction parameter between component elements i and j
E :	Electromotive force
E (superscript):	Excess

T. Matsushita and K. Mukai, *Chemical Thermodynamics in Materials Science*,
https://doi.org/10.1007/978-981-13-0405-7_2

$\mathscr{f}_i$: Fugacity of component i
f_i : Activity coefficient of component i (Raoultian standard state)
f_i^0: Activity coefficient when $x_i \to 0$ (Raoultian standard state, concentration in mole fraction)
$\mathcal{F}$: Number of degrees of freedom of the system (Sect. 7.5)
$\mathcal{F}$: Faraday constant ($= 9.64853 \times 10^4\ \mathrm{C \cdot mol^{-1}}$)
F : Helmholtz energy
G : Gibbs energy
H : Enthalpy
(H) (superscript): Henrian standard state
id (superscript): Ideal
in (subscript): Coming *in*to the system
k : Boltzmann constant
K: Equilibrium constant
$\left(K_{\mathrm{D}}^{\circ}\right)_i$: Partition constant
ln: Natural logarithm
log: Common logarithm
L : Latent heat
L_{ij} : Interaction parameter between atoms i and j
m (subscript): Melting
M: Atomic/molecular weight
Mag (superscript): Magnetic phenomena
[mass% i]: mass% of component i
mix (subscript): Mixing
n: Number of moles
net (subscript): Net amount
out (subscript): Going *out* of the system
P : Pressure
$\mathcal{P}$: Number of phases in the system
Phys (superscript): Physical phenomena
Pres (superscript): Pressure
q Number of independent reactions
Q : Heat
(R) (superscript): Raoultian standard state
re (superscript): Real (non-ideal)
reg (superscript): Regular
rev (subscript): Reversible process
$\mathscr{r}$: Radius of fine particle
R : Gas constant
s: Number of reactions
S: Entropy
sln (superscript): Solution
T : Temperature
U: Internal energy
V : Volume

W : Work

$\mathcal{W}$: The number of possible arrangements (microscopic states) of the system (used in Boltzmann's relation)

x_i : Mole fraction of component i

X_{298}: Quantity X at 298.15 K

$\overline{X}$: Partial molar quantity X

$\overline{X}_i$: Partial molar quantity X of component i

X_i: 1 mol of chemical component i

$\overline{X}^{\mathrm{E}}$: Excess integral molar quantity X (excess molar quantity X of solution)

$\overline{X}_i^{\mathrm{E}}$: Excess partial molar quantity X (excess relative partial molar quantity X)

$\overline{X}_i^{*}$: Quantity X of 1 mol of pure component i

$\Delta_{\mathrm{mix}} X$: Quantity X of mixing

$\Delta_{\mathrm{mix}} \overline{X}$: Molar quantity X of mixing (relative integral molar quantity X)

$\Delta_{\mathrm{mix}} \overline{X}_i$: Relative partial molar quantity X (partial molar quantity X of mixing)

Z : Charge number of ions involved in the cell reaction

Greek alphabet

γ : Heat capacity ratio

γ_i : Activity coefficient of component i (Henrian (1 mass% i) standard state)

γ_i' : Activity coefficient of component i (Henrian standard state)

δ: The meaning of this symbol varies. In some books, it is used to denote a 'virtual variation'; in others, it is an infinitesimal change in non-state functions such as work and heat (δW, δQ. Mathematically, δ is an inexact or imperfect differential. Sometimes the symbol đ (d with a bar) or d' (d-prime) is also used for an infinitesimal change in non-state functions. In this book, only d is used for virtual variation and an infinitesimal change in non-state functions.

Δ (upper-case delta): A 'difference' (or 'change'). For example, Δx is a change in the quantity x. The distance (difference) between position x_1 and x_2 can be described as $\Delta x = x_2 - x_1$ (x_1 is the initial position and x_2 is the final position).

$\varepsilon_i^{(j)}$: Interaction parameter between component elements i and j

η : Efficiency

μ_i : Chemical potential of component i (identical to the partial molar Gibbs energy of component i, $\overline{G}_i$)

$\mu_{\underline{i}}^{\circ(\mathrm{H})}$: Standard chemical potential (Henrian (1 mass% i) standard state)

$\mu_i^{\circ(\mathrm{H})}$: Standard chemical potential (Henrian standard state)

$\mu_i^{\circ(R)}$:	Standard chemical potential (Raoultian standard state)
ν_i and ν_i' :	Number of moles of component i (see Sect. 7.4 for the difference between ν_i and ν_i')
ξ :	Extent of reaction (also known as progress variable or reaction coordinate)
Π:	Osmotic pressure
$\prod_{i=1}^{r}$:	Products (i.e. $\prod_{i=1}^{r} a_i^{\nu_i} = a_1^{\nu_1} \cdot a_2^{\nu_2} \cdot a_3^{\nu_3} \cdots a_r^{\nu_r}$)
σ :	Surface tension
ϕ:	Osmotic coefficient
Ω :	Interaction parameter

Symbols

$\circ$	(superscript): Standard state. For example, μ_i° is the chemical potential at standard state (standard chemical potential)
$*$	(superscript): Pure (property of pure substance). For example, μ_i^* is the chemical potential of pure component i (unary system)
∂:	Partial derivative
$\oint$:	Contour integral
$\int$:	Integral
!:	Factorial

2.2 Glossary

Most of the definitions below are taken from the IUPAC Gold Book (McNaught and Wilkinson 1997) and Kirkwood and Oppenheim's *Chemical Thermodynamics* (1961).

Closed system: In a closed system, no exchange of mass can occur; transfer of heat and work is still possible.

Equilibrium state: See Sect. 2.3.

Extensive property: A thermodynamic property which is dependent on the mass of the system, e.g. Gibbs energy, G; Helmholtz energy, F; enthalpy, H; entropy, S; volume, V; mass, m.

A *system* can be classified based on its intensive properties as being *heterogeneous system* or *homogeneous system*.

Gibbs energy: Gibbs energy, G, is defined as $G = H - TS$, where H is the enthalpy, T is the temperature, and S is the entropy. It has been termed *free energy* or *Gibbs free energy*, but *Gibbs energy* is the term recommended by the IUPAC (see the IUPAC Gold Book).

Heat: The transfer of energy due to a difference in temperature.

Helmholtz energy: Helmholtz energy, F, is defined as $F = U - TS$, where U is the internal energy, T is the temperature, and S is the entropy. It has been termed *free*

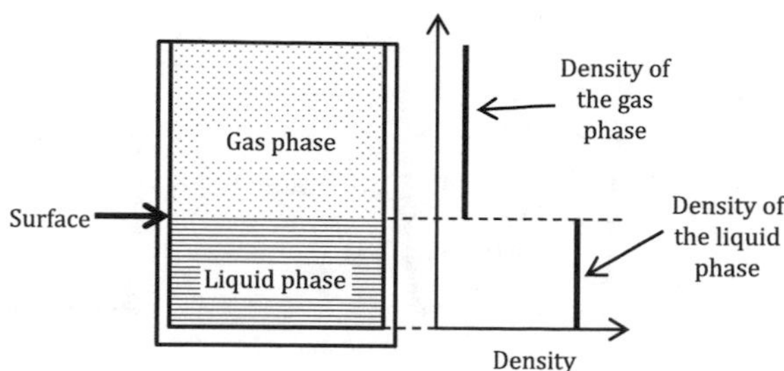

Fig. 2.1 Gas-liquid interface (surface)

energy or *Helmholtz free energy,* but *Helmholtz energy* is the term recommended by the IUPAC (see the IUPAC Gold Book).

Heterogeneous system: A heterogeneous system is one in which intensive properties are not continuous functions of position throughout the system; in other words, a surface or interface in which discontinuities exist. For example, and as is shown in Fig. 2.1, the density of a liquid is discontinuous at its surface. A heterogeneous system is composed of two or more homogeneous regions, each of is termed a *phase*. The number of phases is equal to the number of distinct homogeneous regions.

Homogeneous system: A homogeneous system is one in which intensive properties are continuous functions of position throughout the system. The above definition of a homogeneous system can be re-written in simple terms, assuming the negligible influence of external forces such as gravitational, electrostatic, and magnetic fields: A system that consists of a single phase.

A phase is "a homogeneous part of a heterogeneous system that is separated from other parts by a distinguishable boundary" (Rennie 2016).

However, in the case of the equilibrium conditions of a polymer solution in a centrifugal force field, for example, the concentration is not homogeneous and is greater toward the 'outside' (although the solution as a whole is still regarded as a single phase; Prigogine and Defay 1954). Hence, a general and more precise definition for the term 'phase' is "[a] homogeneous system is one in which the intensive properties are continuous functions of position throughout the system" (Kirkwood and Oppenheim 1961).

Below are examples of heterogeneous system, homogeneous system, and isolated system (Fig. 2.2).

(1) If we define only the liquid phase in Beaker A as a system, it can exchange mass and energy with its surroundings (gas phase and beaker). Hence, the system can be regarded as an open and homogeneous system with a single liquid phase.
(2) If we consider the contents of Beaker B in terms of gas and air phases (but not Beaker B itself) to be a system, only energy can be exchanged and transported. Hence, it is a closed and heterogeneous system which consists of gas, liquid, and solid phases (Beaker A).
(3) Here, there is an insulating container. If we consider its contents in terms of gas and air phases (but not the insulating container itself) to be a system, there is no

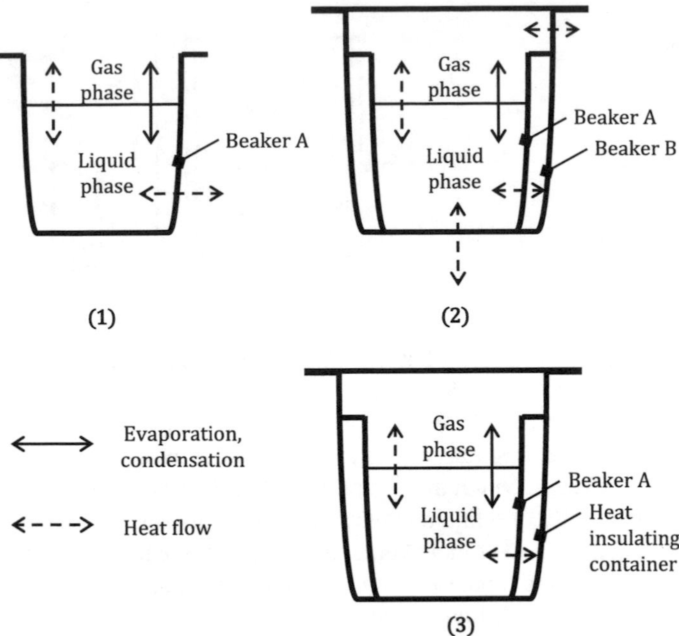

Fig. 2.2 Examples of system classifications

exchange of energy, and it can be regarded as an isolated system which consists of gas, liquid, and solid phases (Beaker A).

Ideal gas: A gas that obeys Boyle's law, and whose energy is a function only of temperature.

Ideal solution: A solution wherein the activity of each component is equal to its mole fraction; a solution that obeys Raoult's law. See also Sects. 8.4.3 and 8.5.4.

Infinitesimal: Vanishingly small, but not zero.

Intensive property: A thermodynamic property that is independent of the mass of the system, e.g. temperature, pressure, concentration, density.

Isolated system: A system that has no interactions with its surroundings. The state of such a system can be described by its macroscopic coordinates, the components of which are termed *thermodynamic properties* or *thermodynamic variables*.

Macroscopic coordinate: A macroscopic coordinate is one whose determination requires only measurements that are averages over regions containing many molecules, over times long with respect to periods of thermal motion, and which involve energies large with respect to individual quanta.

Based on this definition, thermodynamics does not treat molecules, the motion of atoms, structure, and so on individually, but handles averaged properties of overall systems, which can be described by the above-mentioned macroscopic coordinate components.

Thermodynamic properties can be defined as "the properties of the system which describe its macroscopic coordinates", and can be broadly classified into two groups; intensive and extensive properties.

Molar: The adjective 'molar' before the name of an extensive quantity generally means 'division by the amount of substance', i.e. division by the number of moles. The subscript m on an extensive quantity symbol denotes the corresponding molar quantity (see the IUPAC Green Book), e.g. molar enthalpy, $H_\mathrm{m} = H/n$.

In this book, molar quantities are represented by a bar over the quantity, as is recommended by the IUPAC (e.g. $\overline{H}$ is used instead of H_m).

Non-ideal gas: A 'real' gas that does not show ideal behaviour. See also 'Ideal gas'.

Non-ideal solution: A 'real' solution that does not show ideal behaviour, i.e. $\Delta_\mathrm{mix} H$ is not equal to zero, and $\Delta_\mathrm{mix} S$ is not the same as for an ideal solution (see Sects. 8.4.3 and 8.5.4). See also 'Ideal solution'.

Open system: A system that exchanges mass, heat, and work with its surroundings.

Real gas: See 'Non-ideal gas'.

Real solution: See 'Non-ideal solution'.

Reference state (of an element): The state in which an element is stable at a given standard-state pressure and temperature, the values for which are not specified. In the thermodynamic software package Thermo-Calc, 'standard element reference' (SER) refers to an element in its most stable form at a temperature of 298.15 K and pressure of 1 bar.

Regular solution: A solution for which $\Delta_\mathrm{mix} S$ is the same as for an ideal solution, but $\Delta_\mathrm{mix} H$ is not (i.e. $\Delta_\mathrm{mix} H$ is not zero) (Sects. 8.4.3 and 8.5.4).

Specific: The adjective 'specific' before the name of an extensive quantity denotes a physical quantity obtained by division by mass. When the symbol for extensive quantity is a capital letter, the symbol used for the specific quantity is often the corresponding lower-case letter, e.g. specific heat capacity at constant pressure, $c_P = C_P/m$, where m is the mass.

Standard chemical potential: The standard chemical potential of Substance B at temperature T, $\mu^\circ_\mathrm{B}(T)$, is the value of the chemical potential under given standard conditions (in a standard state). See also 'Standard thermodynamic quantities'.

Standard condition: This term is not defined in the IUPAC Gold Book and is here used in a relatively general sense, with no set definition. It should not be confused with the technical term 'standard conditions for gases'.

Standard conditions for gases: A set of conditions that allow the results of experimental measurements to be compared; a temperature of 273.15 K (0 °C) and pressure of 10^5 pascals. IUPAC recommends that the use of 1 atm as standard pressure (equivalent to 1.01325×10^5 Pa) should be discontinued, and so is not used in this book. 'Standard conditions for gases' should not be confused with the general term 'standard condition', which is also used in this book.

Standard pressure: A given value of pressure, denoted by $\mathrm{p}^\ominus$ or p°. In 1982, IUPAC recommended a value of 10^5 Pa; prior to 1982, a value of 101325 Pa (=1 atm) was usually used. In this book, P is used instead of p for pressure, which is equally acceptable according to IUPAC. The difference between 1 atm and 1 bar is generally considered to be negligible in practice due to the uncertainty of available data.

Standard reaction quantities: Infinitesimal changes in thermodynamic functions with *extent of reaction* divided by the infinitesimal increase in the extent when all of the reactants and products are in their standard states. For the quantity X this should be denoted by $\Delta_r X^\circ$, but usually only ΔX° is used. For specific types of reaction the subscript r is replaced by f for formation, c for combustion, a for atomisation, and superscripted ‡ for activation, e.g. standard reaction enthalpy, $\Delta_r H^\circ$.

Standard state: The state of a system chosen as standard for reference by convention. In the field of materials science, the following definition is conventionally used: "The standard state is customarily chosen to be at a pressure of 1 bar and in the most stable structure of that element at the temperature at which it is investigated." The standard state of liquids and solids is the state of the pure substance when subjected to a total pressure of 1 bar. Thus, temperature is not part of the definition of a standard state. However, in most thermodynamics data books, the properties of components at 298.15 K (25 °C) are compiled. To denote the standard state, $\circ$ (a superscripted circle) is commonly used.

Standard state of chemical potential: In the IUPAC Green and Gold Books, the following standard states are described:

For a gas phase:

"It is the (hypothetical) state of the pure substance in the gaseous phase at the standard pressure $P = P^\circ$, assuming ideal behaviour."

This definition corresponds to $\mu_i^{\circ,\mathrm{id(g)}}$ in Eq. (8.5) in this book.

For a pure phase, mixture, or solvent in a liquid or solid state:

"It is the state of the pure substance in the liquid or solid phase at the standard pressure, $P = P^\circ$."

This definition corresponds to $\mu_i^{\circ,(\ell)}$ in Eq. (8.21) in this book.

For a solute in solution:

"It is the (hypothetical) state of solute at the standard molality m°, standard pressure P°, or standard concentration c° and exhibiting infinitely dilute solution behaviour."

This definition corresponds to e.g. $\mu_i^{\circ(\mathrm{H})}$ in Eq. (8.61) in this book.

STP: An abbreviation that stands for 'standard temperature and pressure' (273.15 K or 0 °C and 10^5 Pa, respectively), usually used to report gas volumes. Note that flow meters calibrated at standard gas volumes per unit time often refer to volumes at 25 °C, not 0 °C. This term is not used in this book, but should not be confused with the term 'standard state'.

Standard thermodynamic quantities: Values of thermodynamic functions in a standard state, characterised by standard pressure, molality, or amount concentration, but not by temperature. Standard quantities are denoted by adding the superscript $\ominus$ or $\circ$ to the symbol of the quantity. In this book, standard quantities are denoted by a superscripted $\circ$, e.g. standard chemical potential, μ_i°.

State function (also known as thermodynamic function): An amount that depends only on the state of the system, and not on the path by which the system reached its current state.

Surroundings: The rest of the physical world. Thermodynamic systems are categorised based on how they interact with their surroundings.

System: The part of the physical world that is under consideration. According to this definition, a system can be chosen arbitrarily so that it becomes easy to deal with, e.g. a water solution in a beaker or metal/slag in a crucible. However, as is discussed in the definition of macroscopic coordinates, the size of a system must exceed a certain size.

Temperature: Thermodynamic temperature. Absolute temperature is used in this book, the unit of which is Kelvin.

Thermodynamic function: See 'State function'.

Work: The mechanical energy transferred by a system to its surroundings, or vice versa.

2.3 Equilibrium State

What is an 'equilibrium state'? In this book we focus on 'thermodynamic equilibrium states', as the study of these allows for a deeper understanding of chemical thermodynamics.

The thermodynamic state of a system is defined by the combination of several intensive properties. In a thermodynamic equilibrium state, both Conditions (a) and (b) are satisfied:

(a) Intensive properties are not dependent on time.
(b) There is no exchange of energy and mass inside the system and at the interface with its surroundings.

When only (a) is satisfied, a system is in a steady state. Consider a system that consists of a solid material (Fig. 2.3): the left and right sides have been maintained at T_1 and T_2 respectively for a sufficient period of time. Here, Condition (a) is satisfied but (b) is not, as there is heat flux in the solid phase and heat exchange between the solid phase and the surroundings. Hence, this system is in a steady state, rather than a thermodynamic equilibrium state.

It is important to understand the concept of the equilibrium state—so, let us discuss it in more detail.

Mechanical equilibrium: A physical system that consists of several objects is in mechanical equilibrium if the net force (sum of external forces) and interaction between objects is zero.

In Fig. 2.4, a weight hangs from a spring. When it is at rest, the spring-weight system is in a mechanical equilibrium state under the given conditions.

Thermal equilibrium: The equilibrium state for heat phenomena can be treated in a similar manner, and so thermal equilibrium can be described as "[a] state in

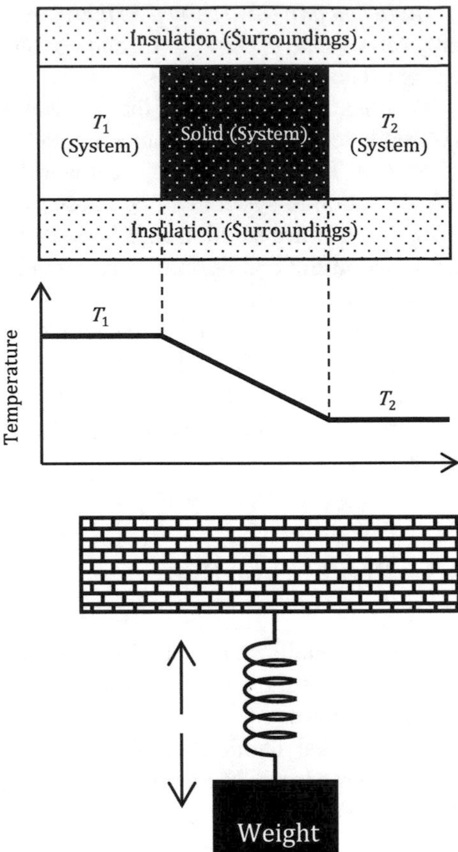

Fig. 2.3 Steady state

Fig. 2.4 Mechanical equilibrium

which two objects, or an object and its surroundings, have the same temperature so that there is no exchange of heat energy between them" (Ridpath 2012). In Fig. 2.5, a glass ball hanging from a string is submerged in water in a sealed thermostatic chamber. If this system is left for long enough, the temperatures of the ball and the water will become the same, meaning that no heat transfer occurs between the ball and the water. In such a case, the ball-water system is in a thermal equilibrium state.

Thermodynamic equilibrium: As is discussed above, thermodynamic equilibrium, which involves energy (heat, work) transfer and mass transfer, is more complex than mechanical equilibrium and thermal equilibrium. Let us create a gas phase by removing some of the water from the sealed thermostatic chamber in Fig. 2.5 and setting the temperature of the thermostat to T_1. Heat transfer will then occur between the thermostat and the water, and the temperature of the water will thus change. Furthermore, a change in the water vapour pressure will occur. Thus, mass exchange occurs between the gas and liquid phases, and the generation of heat of evaporation or heat of condensation occurs at the same time as the temperature of the

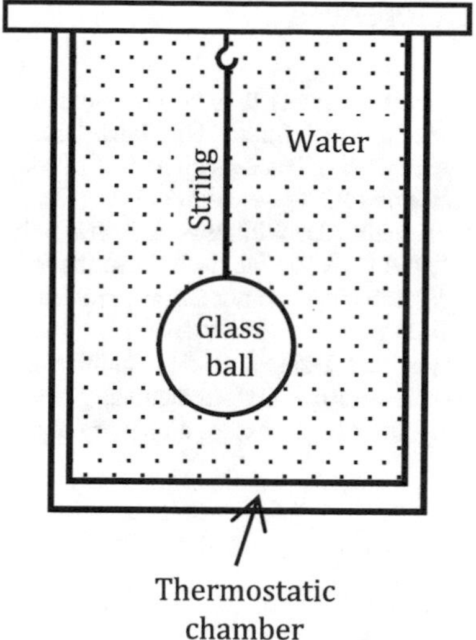

Fig. 2.5 Thermal equilibrium

system is changed. With these complex changes and after a certain period of time, the temperature, T_1, and water vapour pressure corresponding to this temperature will become constant, and the system will reach a thermodynamic equilibrium state. With new conditions, achieved by e.g. setting the temperature of the thermostat to T_2, the conditions of the system will change and the system will reach a new thermodynamic equilibrium state corresponding to the new temperature, T_2.

How many thermodynamic properties are required to describe the thermodynamic equilibrium state of a system? Based on experiments, it is known that $r + 1$ intensive properties are enough for a homogeneous open system that consists of r independent components, where the number of independent components can be defined as "the minimum number of chemical substances from which the system can be prepared by a specified set of procedures" (Kirkwood and Oppenheim 1961, p. 2). This definition, however, is difficult to understand, and is discussed in the chapter on the phase rule (Sect. 7.5). More precisely, the following condition should be satisfied: "No external electric, magnetic, or gravitational fields and that the only force which may act on the system is a uniform normal pressure" (Kirkwood and Oppenheim 1961, p.2). In a general reaction system, this condition is satisfied in most cases.

The most commonly used intensive properties are T, P and (for defining composition) $r - 1$ variables. Hence, the intensive properties of a system are functions of T, P, $x_1, x_2 \cdots x_{r-1}$. *Extensive properties* are functions of T, P,$m_1, m_2 \cdots m_r$ (where m_i is the mass of the i-th component), and are thus a function of $r + 2$ variables.

References

Cohen ER, Cvitas T, Frey JG, Holmstrom B, Kuchitsu K, Marquardt R, Mills I, Pavese F, Quack M, Stohner J, Strauss HL, Takami M, Thor AJ, Quantities (2008) Units and symbols in physical chemistry, IUPAC green book, 3rd edn, 2nd Printing. IUPAC and RSC Publishing, Cambridge

Kirkwood JG, Oppenheim I (1961) Chemical thermodynamics, 1st edn. McGraw-Hill

McNaught AD, Wilkinson A (1997) IUPAC. Compendium of chemical terminology, 2nd ed. (the "gold book"). Blackwell Scientific Publications, Oxford (XML on-line corrected version: http://goldbook.iupac.org (2006-) created by Nic M, Jirat J, Kosata B; updates compiled by Jenkins AD)

Prigogine I, Defay R (1954) Chemical thermodynamics. Longmans, Green and Co Ltd

Rennie R (2016) Oxford dictionary of chemistry, 7th edn. Oxford University Press

Ridpath I (2012) Oxford dictionary of astronomy, 2nd rev edn. Oxford University Press

Chapter 3
The First Law of Thermodynamics

In this chapter, the first law of thermodynamics—the relationship between internal energy, heat, and work—is discussed. The first law of thermodynamics will be used to e.g. derive the criteria for equilibrium (see Chap. 7).

3.1 The First Law of Thermodynamics

Mechanical energy transfer is a work process, and thermal energy transfer is a thermal process. As is shown in Fig. 3.1, with the amount of energy transferred mechanically (the *work* done by the system towards its surroundings, W_{out}) and the amount of energy transferred thermally (the absorbed *heat* from the surroundings by the system, Q_{in}), the change in the *internal energy* (hereafter simply *internal energy change* and the same for the other properties) of the closed system, ΔU (where Δ denotes finite displacement), is described by:

$$\Delta U = Q_{\mathrm{in}} - W_{\mathrm{out}} \tag{3.1}$$

Equation (3.1) is a mathematical expression of the *first law of thermodynamics*.

Here, Q_{in} and W_{out} may be positive or negative (or zero). A positive Q_{in} value represents heat being added to the system, and a positive W_{out} value represents the work done by the system. The signs for heat and work differ between books (in Prigogine and Defay (1954), for example, the work done *to* the system is taken as positive), and so it is recommended that the signs are always carefully defined.

3.2 Internal Energy, *U*

U is *internal energy*. According to the first law of thermodynamics, ΔU does not depend on the path of the change in the case of a closed system, but does depend on

T. Matsushita and K. Mukai, *Chemical Thermodynamics in Materials Science*,
https://doi.org/10.1007/978-981-13-0405-7_3

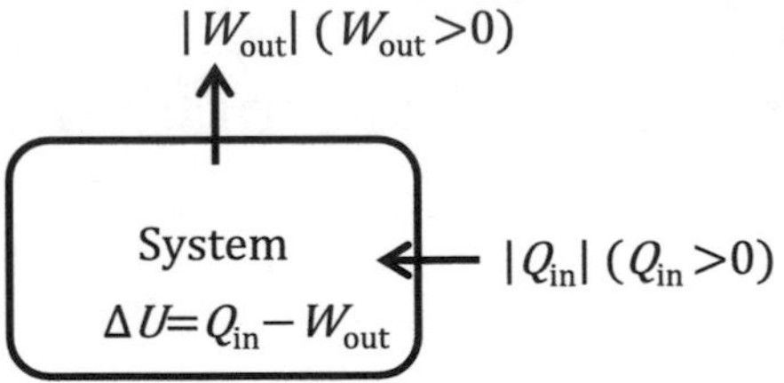
$|W_{out}|$ ($W_{out}>0$)
System
$\Delta U = Q_{in} - W_{out}$
$|Q_{in}|$ ($Q_{in}>0$)
$Q_{in}>0$: The heat $|Q_{in}|$ is added to the system
$W_{out}>0$: The work $|W_{out}|$ is done *by* the system

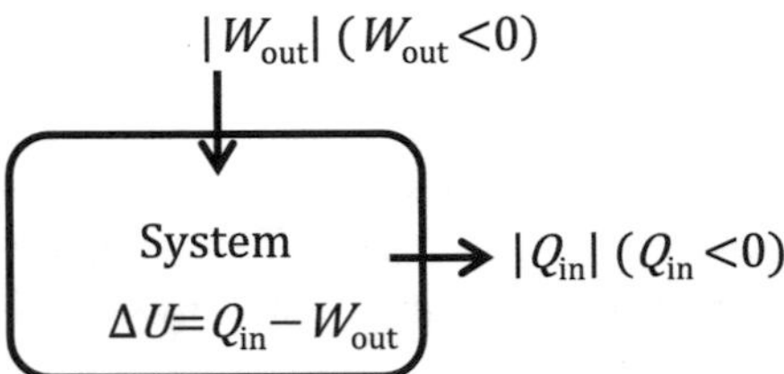
$|W_{out}|$ ($W_{out}<0$)
System
$\Delta U = Q_{in} - W_{out}$
$|Q_{in}|$ ($Q_{in}<0$)
$Q_{in}<0$: The heat $|Q_{in}|$ is released from the system
$W_{out}<0$: The work $|W_{out}|$ is done *on* the system

Fig. 3.1 The relationship between ΔU, Q_{in}, and W_{out}

the initial and final conditions of the system. U also depends solely on the state of the system, and is categorised as a *state function* (also known as the *thermodynamic function*). Temperature (T), pressure (P), volume (V), entropy (S), enthalpy (H), Helmholtz energy (F), and Gibbs energy (G) are also state functions.

The following example clearly demonstrates that internal energy, U, is a state function. As is shown in Fig. 3.2, the internal energy change is ΔU_a when the state of the system changes from State I to State II via Path a, and the internal energy change is ΔU_b when the state of the system changes from State II to State I via Path b. According to the first law of thermodynamics, the total energy inside and outside the system is conservative. Hence, when the system changes via $\mathrm{I} \overset{a}{\rightarrow} \mathrm{II} \overset{b}{\rightarrow} \mathrm{I}$, the change in the system is $(\Delta U_a + \Delta U_b)$ and the system releases the energy of $-(\Delta U_a + \Delta U_b)$ to the surroundings. Thus, if $\Delta U_a + \Delta U_b < 0$ $(U_1' - U_1 < 0)$ and the value of U is different for the same State I, the surroundings can perpetually take energy from the closed system as a result of the cycle of $\mathrm{I} \overset{a}{\rightarrow} \mathrm{II} \overset{b}{\rightarrow} \mathrm{I}$. Of course, this violates the first law of thermodynamics, which denies the existence of a *perpetual motion machine of the first kind* (a machine that can perpetually produce work without energy input from its surroundings).

To satisfy the first law of thermodynamics for this kind of cycle, it must be the case that $\Delta U_a + \Delta U_b = 0$, i.e. $U_1' = U_1$. This is just an example, but shows that U is a function which is defined by the state, and does not depend on the path of the change.

Fig. 3.2 Internal energy change

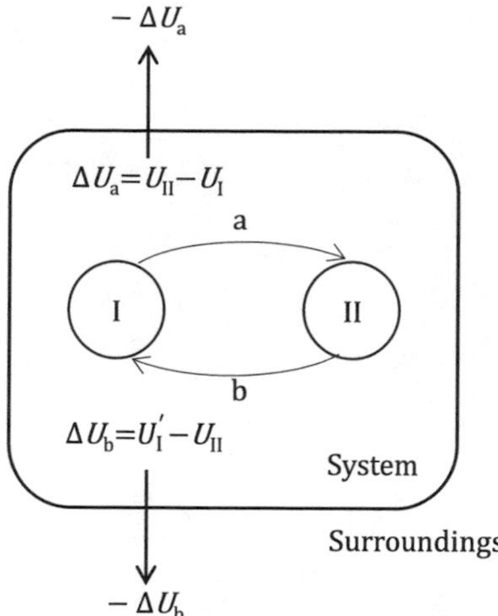

Let us now explore whether the functions Q_{in} and W_{out} in the equation $\Delta U = Q_{\mathrm{in}} - W_{\mathrm{out}}$ are also state functions. As can be seen in Example 3.1 below, when the state of an ideal gas is changed under constant temperature, the values of Q_{in} and W_{out} depend on the path of the change—which is to say that these functions have different values for reversible expansion and expansion under constant external pressure. From this fact, one can understand that Q_{in} and W_{out} are not state functions.

Example 3.1

Calculate Q_{in} and W_{out} when 1 mol of an ideal gas at 3×10^5 Pa expands at 323 K (50 °C). The pressure after the expansion is 10^5 Pa. Calculate these values for (1) reversible expansion and (2) expansion under constant external pressure.

Solutions:

(1) The internal energy of an ideal gas depends solely on temperature. Hence, $\Delta U = 0$ for both (1) and (2) (at a constant temperature).

$$W_{\mathrm{out}} = \int_{V_{\mathrm{i}}}^{V_{\mathrm{f}}} P\mathrm{d}V$$

$$P = RT/V$$

where P is the pressure, V is the volume, V_{i} is the initial volume, V_{f} is the final volume, R is the gas constant ($=8.314\ \mathrm{J \cdot mol^{-1} \cdot K^{-1}}$), and T is the temperature.

Hence,

$$W_{\text{out}} = RT \int_{V_i}^{V_f} \left(\frac{dV}{V}\right) = RT \ln\left(\frac{V_f}{V_i}\right) dV = RT \ln\left(\frac{P_i}{P_f}\right)$$

$$= 8.314 \times 323 \times \ln\left(\frac{3 \times 10^5}{10^5}\right) = 2950\text{J}$$

where P_i and P_f are the pressure in the initial and final states, respectively. According to the first law of thermodynamics, $\Delta U = Q_{in} - W_{out} = 0$ at a constant temperature.

Hence, $W_{out} = Q_{in} = 2950$ J.

(2) As shown in Fig. 3.3, when an ideal gas is expanded under external pressure, $P_{ex} = 10^5$ Pa,

$$\begin{aligned} W_{\text{out}} &= \int_{V_i}^{V_f} P_{ex} dV \\ &= P_{ex}(V_f - V_i) \\ &= RT P_{ex}\left(\frac{1}{P_f} - \frac{1}{P_i}\right) \\ &= 1790\text{J} = Q_{in} \end{aligned}$$

As can be seen from these results, the values for Q_{in} and W_{out} depend on the path of the change.

However, under the following conditions, the values of non-state functions (heat, work) become identical to the change in state function (ΔU):

(a) Adiabatic change
$\Delta U = -W_{out}$ as $Q_{in} = 0$ (adiabatic system), and W_{out} becomes equal to the internal energy change, ΔU.

(b) When only uniform normal pressure is exerted on the system

$$W_{\text{out}} = P\Delta V \tag{3.2}$$

where ΔV is the *volume change*.

Hence, at a constant volume ($\Delta V = 0$), $W_{out} = 0$.

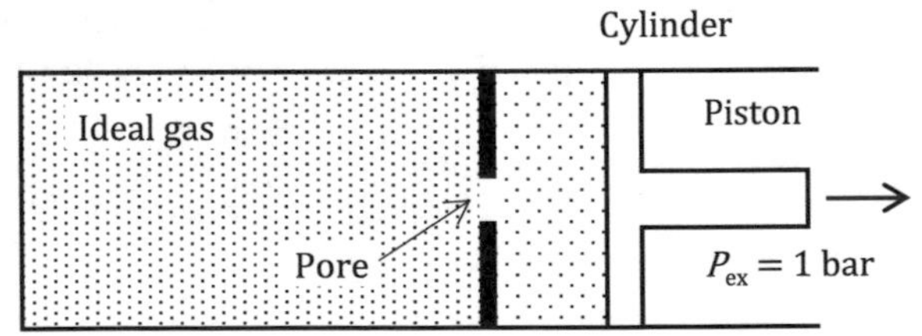

Fig. 3.3 Expansion of an ideal gas at constant external pressure

Hence,

$$\Delta U = Q_{\text{in},V} \tag{3.3}$$

The heat absorbed by the system at a constant volume, $Q_{\text{in},V}$, is equal to the internal energy change, ΔU (state function).

Under constant pressure,

$$\Delta U = Q_{\text{in},P} - P\Delta V \tag{3.4}$$

where $Q_{\text{in},P}$ is the heat absorbed by the system at a constant pressure.

Hence,

$$Q_{\text{in},P} = \Delta U + P\Delta V \tag{3.5}$$

The value of $Q_{\text{in},P}$ is also independent of the path of the change, as both P and V are state functions. $Q_{\text{in},P}$ is identical to change in enthalpy, as is discussed in Sect. 4.1, Eq. (4.3).

Under the conditions described above one can treat W_{out} and Q_{in} as state functions, as these values can be determined based on the state of the system and are measurable values. Hence, W_{out} and Q_{in} are useful and important values in practice.

3.3 The Nature of Internal Energy

Q_{in} and W_{out} are produced by the physical changes of a system, i.e. temperature change (Q_{in}), changes in state such as melting and evaporation (Q_{in}) and expansion and compression (W_{out}), and the chemical changes of a system such as chemical reactions (Q_{in}). The Q_{in} and W_{out} values resulting from macroscopic changes such as physical and chemical changes can be interpreted from a microscopic perspective, as well based on the kinetic energy of the atoms and molecules of a system (translational, rotational, and vibrational kinetic energies) and changes in the interaction energies between electrons, atoms, and molecules. This is to say that the nature of internal energy relates to the above-discussed kinetic and interaction energies of atoms and molecules. Hence, although it is possible to calculate the absolute value of internal energy in principle and from a structural point of view, this is extremely difficult in practice. In most cases, however, relative internal energy change, ΔU, is much more important in thermodynamics than absolute values of internal energy, U.

It should also be noted that nuclear energy (mc^2, where m is the mass of the whole system and c is the speed of light) and mass energy are not included in internal energy. Mass energy is the kinetic energy ($\frac{1}{2}mv^2$, where v is the velocity) and potential energy ($mg\Delta h$, where g is the gravitational acceleration and Δh is the change in height) of the whole system.

3.4 Summary

The first law of thermodynamics:
The first law of thermodynamics can be mathematically described as

$$\Delta U = Q_{\text{in}} - W_{\text{out}} \tag{3.1}$$

where ΔU is the internal energy change, a positive Q_{in} value is the heat added to a system, a negative Q_{in} value is the heat released from a system, a positive W_{out} value is the work done by the system, and a negative W_{out} value is the work done to the system.

State function:
A state function is a function that is defined for state variables or state quantities, is dependent only on the state of the system, and does not depend on the path by which the system reached its current state. Internal energy, U, is a state function, but heat, Q_{in}, and work, W_{out}, are not.

Reference

Prigogine I, Defay R (1954) Chemical thermodynamics. Longmans, Green and Co Ltd

Chapter 4
Enthalpy, *H*

In this chapter we define a new state function—enthalpy, H—which is used to express the heat absorbed by a system under constant pressure. After a general discussion of the heat of reaction/formation, two important laws—(1) Kirchhoff's law of thermochemistry (which describes the influence of temperature on the heat of reaction) and (2) Hess's law (which is used to calculate the unknown heat of reaction based on the known heat of reaction)—are discussed.

4.1 Enthalpy, *H*, and Heat Capacities

Enthalpy, H, is defined by the following equation:

$$H \equiv U + PV \tag{4.1}$$

where U is the internal energy, P is the equilibrium pressure of the system, and V is the volume. The unit for enthalpy is that of energy, J.

According to the first law of thermodynamics, in the case of an isochoric (constant-volume; $\Delta V = 0$, therefore $W_{\text{out}} = P\Delta V = 0$) process under only uniform normal pressure conditions, the heat absorbed by a system, $Q_{\text{in},V}$, is described as follows:

$$Q_{\text{in},V} = \Delta U = U_2(T_2, V) - U_1(T_1, V) \tag{4.2}$$

Thus, the heat absorbed by the system ($Q_{\text{in},V}$) is equal to the increase in the internal energy of the system (ΔU).

On the other hand, in the case of an isobaric (constant-pressure) process, the heat absorbed by the system, $Q_{\text{in},P}$, is described as follows:

T. Matsushita and K. Mukai, *Chemical Thermodynamics in Materials Science*,
https://doi.org/10.1007/978-981-13-0405-7_4

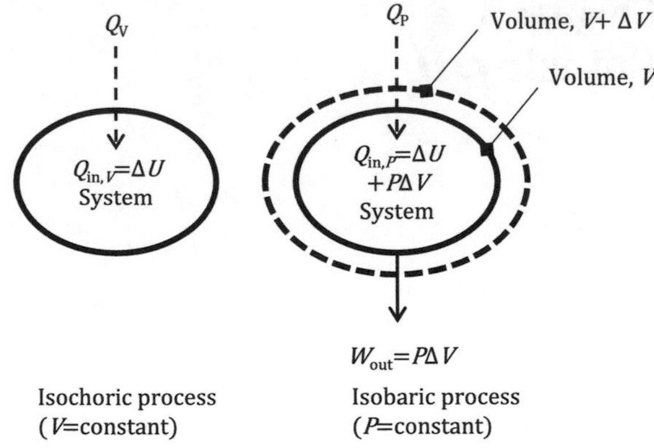

Fig. 4.1 The relationship between $Q_{\text{in},P}$, ΔU, and $P\Delta V$ for isochoric and isobaric processes

$$\begin{aligned} Q_{\text{in},P} &= U_2(T_2, V_2) - U_1(T_1, V_1) + P(V_2 - V_1) \\ &= (U_2 + PV_2) - (U_1 + PV_1) \\ &= \Delta H \end{aligned} \tag{4.3}$$

As can be seen from this derivation, $Q_{\text{in},P}$ is equal to the *enthalpy change*, ΔH. As pressure is kept constant in most chemical reaction systems, isobaric processes can be considered to be important and $Q_{\text{in},P} = \Delta H$ to be a useful relationship.

As shown in Fig. 4.1, $Q_{\text{in},P}$ consists of ΔU and the work done by a system to its surroundings, W_{out}.

The heat that is produced or absorbed by a system is equal to the enthalpy change in the case of an isobaric (constant-pressure) process. Hence, it is possible to obtain the heat balance of the system by calculating the enthalpy change when the pressure of the system is constant.

Based on the discussion of isochoric and isobaric processes, two heat capacities can be defined.

Heat capacity at constant volume, C_V:

$$C_V = \left(\frac{\partial Q}{\partial T}\right)_V = \left(\frac{\partial U}{\partial T}\right)_V \tag{4.4}$$

Heat capacity at constant pressure, C_P:

$$C_P = \left(\frac{\partial Q}{\partial T}\right)_P = \left(\frac{\partial H}{\partial T}\right)_P \tag{4.5}$$

As can be seen, heat capacity is the heat required to increase the temperature by an amount, $\mathrm{d}T$, the unit of which is $\mathrm{J \cdot K^{-1}}$.

Heat capacity at a constant volume per unit mass is termed *specific heat capacity at constant volume,* c_V, and heat capacity at constant pressure per unit mass is *specific heat capacity at constant pressure,* c_P. The unit for both is $\mathrm{J \cdot kg^{-1} \cdot K^{-1}}$.

Heat capacity at a constant volume per mole and constant pressure per mole can be also defined. In this case, the unit is $\text{J} \cdot \text{mol}^{-1} \cdot \text{K}^{-1}$. These quantities are termed *molar heat capacity at constant volume,* $\overline{C}_V$, and *molar heat capacity at constant pressure,* $\overline{C}_P$, respectively.

For usage of the adjectives 'specific' and 'molar', see their definitions (Sect. 2.2).

Now, we will derive the relationship between C_V and C_P for 1 mol of an ideal gas. In the case of a closed system,

$$U = U(T, V) \tag{4.6}$$

and the total derivative of U is

$$\mathrm{d}U = \left(\frac{\partial U}{\partial T}\right)_V \mathrm{d}T + \left(\frac{\partial U}{\partial V}\right)_T \mathrm{d}V \tag{4.7}$$

In the case of an isobaric (constant-pressure) process, its partial derivative becomes

$$\left(\frac{\partial U}{\partial T}\right)_P = \left(\frac{\partial U}{\partial T}\right)_V + \left(\frac{\partial U}{\partial V}\right)_T \left(\frac{\partial V}{\partial T}\right)_P \tag{4.8}$$

The internal energy of an ideal gas is a function only of temperature (this is a definition of an ideal gas). Therefore, $\left(\frac{\partial U}{\partial V}\right)_T = 0$ in Eq. (4.8). Using Eqs. (4.4) and (4.8), in the case of an ideal gas,

$$\left(\frac{\partial U}{\partial T}\right)_P = \left(\frac{\partial U}{\partial T}\right)_V = C_V \tag{4.9}$$

According to Eqs. (4.1) and (4.5),

$$C_P = \left(\frac{\partial H}{\partial T}\right)_P = \left(\frac{\partial U}{\partial T}\right)_P + \left(\frac{\partial (PV)}{\partial T}\right)_P \tag{4.10}$$

Based on Eqs. (4.9) and (4.10),

$$C_P = C_V + P\left(\frac{\partial V}{\partial T}\right)_P \tag{4.11}$$

We are considering 1 mol of an ideal gas and, according to the ideal gas law, $V = RT/P$ (where R is the gas constant). Thus,

$$\left(\frac{\partial V}{\partial T}\right)_P = \frac{R}{P} \tag{4.12}$$

Hence,

$$C_P = C_V + R \tag{4.13}$$

Equation (4.13) is a relationship between heat capacity at a constant pressure, C_P, and heat capacity at a constant volume, C_V.

4.2 Definition of Enthalpy, $H = 0$

Obtaining an absolute value for enthalpy is near-impossible due to the fact that enthalpy includes internal energy, (U). However, by specifying a reference, the relative difference between enthalpy and the reference can be calculated. For the sake of convenience, the enthalpy of elements in most stable states, i.e. in a reference state, at under 1 bar, i.e. in a standard state, and 298.15 K (25 °C) can be defined as zero, $H^\circ_{i,298} \equiv 0$ (the superscript o denotes the amount of the thermodynamic property in a standard state). Note that it is defined in this way in order to facilitate obtaining the 'absolute value' of enthalpy, making calculation, tabulation, etc. easier, but the actual value of $H^\circ_{i,298}$ is not zero.

4.3 Heat of Reaction, $\Delta_r H$

During a chemical reaction, heat is generated or absorbed by a system. The exchange of heat between a reaction system and its surroundings as a result of a chemical reaction is termed the *heat of reaction*. For a reaction under constant temperature and pressure conditions, the heat of reaction is equal to the enthalpy change, $\Delta_r H$. Note that when $\Delta_r H > 0$, an endothermic reaction is occurring (heat is absorbed by the system), and when $\Delta_r H < 0$, an exothermic reaction is occurring (heat is generated by the system, and the system releases the heat to the surroundings) as enthalpy is considered from the viewpoint of the system.

Consider the following reaction:

$$\nu_1' X_1 + \nu_2' X_2 = \nu_3' X_3 + \nu_4' X_4 \tag{4.14}$$

where ν_i' is the number of moles of component i and X_i is 1 mol of the chemical component i.

The $\Delta_r H$ of the above reaction is

$$\Delta_r H = \nu_3' \overline{H}_3 + \nu_4' \overline{H}_4 - \left(\nu_1' \overline{H}_1 + \nu_2' \overline{H}_2\right) \tag{4.15}$$

where $\overline{H}_i$ is the *partial molar enthalpy* of component i (See Chap. 9).

The term 'heat of reaction', $\Delta_r H$, changes based on the type of reaction, and is termed

heat of formation for the formation of a compound from its elements (e.g. $2Al + \frac{3}{2}O_2 \rightarrow Al_2O_3$).

heat of combustion for combustion (e.g. $C + O_2 \rightarrow CO_2$).
heat of decomposition for decomposition (e.g. $2CO \rightarrow C + CO_2$).
heat of melting (or *heat of fusion*) for melting.
heat of vapourisation for vapourisation.
heat of solution for solution (e.g. $Si(\ell) \rightarrow \underline{Si}$(dissolved in molten iron)).

4.4 Influence of Temperature on the Heat of Reaction (Kirchhoff's Law)

The heat of reaction, $\Delta_r H$, depends on temperature, and can be expressed as a function of heat capacity.

As is described in the previous section, $\Delta_r H$ of the chemical reaction

$$\nu_1' X_1 + \nu_2' X_2 = \nu_3' X_3 + \nu_4' X_4 \tag{4.16}$$

is

$$\Delta_r H = \nu_3' \overline{H}_3 + \nu_4' \overline{H}_4 - \left(\nu_1' \overline{H}_1 + \nu_2' \overline{H}_2\right) \tag{4.17}$$

As we usually consider a reaction to occur under a constant pressure, differentiating Eq. (4.17) with respect to temperature, T, obtains the following relationships:

$$\left(\frac{\partial \Delta_r H}{\partial T}\right)_P = \nu_3' \left(\frac{\partial \overline{H}_3}{\partial T}\right)_P + \nu_4' \left(\frac{\partial \overline{H}_4}{\partial T}\right)_P - \nu_1' \left(\frac{\partial \overline{H}_1}{\partial T}\right)_P - \nu_2' \left(\frac{\partial \overline{H}_2}{\partial T}\right)_P \tag{4.18}$$

According to Eq. (4.5),

$$\left(\frac{\partial \overline{H}_i}{\partial T}\right)_P = \overline{C}_{P,i} \tag{4.19}$$

Hence,

$$\begin{aligned}\left(\frac{\partial \Delta_r H}{\partial T}\right)_P &= \nu_3' \overline{C}_{P,3} + \nu_4' \overline{C}_{P,4} - \left(\nu_1' \overline{C}_{P,1} + \nu_2' \overline{C}_{P,2}\right) \\ &= \Delta C_P \end{aligned} \tag{4.20}$$

This relationship is true not only for the reaction given in Eq. (4.16), but for all chemical reactions. It can be expressed in a general form as follows:

$$\left(\frac{\partial \Delta H}{\partial T}\right)_P = \Delta C_P \tag{4.21}$$

As is discussed in the section on chemical equilibrium (Sect. 7.4), it can be easily understood that the above equation is true for all chemical reactions by considering the generalised chemical reaction formula. Equation (4.21) is called *Kirchhoff's law*.

By integrating Eq. (4.21), the following equation can be obtained:

$$\Delta H = \int \Delta C_P \mathrm{d}T + \text{Constant} \tag{4.22}$$

4.5 Temperature Dependence of Enthalpy, *H*

Enthalpy, H, is an extensive property and a function of temperature, pressure, and number of moles. Hence, it may be described as $H(T, P, n)$, where n is the number of moles. Enthalpy per mole ($\overline{H} = H/n$) is described as $\overline{H}(T, P)$, and its total derivative is:

$$\mathrm{d}\overline{H} = \left(\frac{\partial \overline{H}}{\partial T}\right)_P \mathrm{d}T + \left(\frac{\partial \overline{H}}{\partial P}\right)_T \mathrm{d}P \tag{4.23}$$

In the case of an isobaric (constant-pressure) process,

$$\mathrm{d}\overline{H} = \left(\frac{\partial \overline{H}}{\partial T}\right)_P \mathrm{d}T = \frac{1}{n} C_P \mathrm{d}T \tag{4.24}$$

because $\left(\frac{\partial \overline{H}}{\partial T}\right)_P = \frac{1}{n} C_P$ (cf. Eq. (4.5)).

The change in molar enthalpy, $\Delta\overline{H}$, as a result of the temperature changing from T_1 to T_2 is

$$\begin{aligned}\Delta\overline{H} &= \overline{H}(T_2, P) - \overline{H}(T_1, P)\\ &= \frac{1}{n}\int_{T_1}^{T_2} C_P \mathrm{d}T = \int_{T_1}^{T_2} \overline{C}_P \mathrm{d}T\end{aligned} \tag{4.25}$$

If the relationship between heat capacity at a constant pressure per mole—i.e. molar heat capacity at constant pressure, $\overline{C}_P$—and temperature is obtained from a data book, the enthalpy change as a result of the change in temperature of the unary system, $\Delta\overline{H}$, can be obtained by performing the integration in Eq. (4.25).

As materials can change state (e.g. Solid ↔ Liquid ↔ Gas), *latent heat* (the heat released or absorbed by a system), which relates to phase transformations, must be considered in order to calculate $\Delta\overline{H}$. For example, when the solid phase transformation I → II occurs at T_t, melting occurs at T_m, and evaporation occurs at T_e between the temperatures T_1 and T_2, $\Delta\overline{H}$ can be expressed as follows:

$$\Delta\overline{H} = \frac{1}{n}\left[\int_{T_1}^{T_t} C_P^{(S_I)}\mathrm{d}T + L_t + \int_{T_t}^{T_m} C_P^{(S_{II})}\mathrm{d}T + L_m + \int_{T_m}^{T_e} C_P^{(\ell)}\mathrm{d}T + L_e + \int_{T_e}^{T_2} C_P^{(g)}\mathrm{d}T\right] \tag{4.26}$$

where S_I and S_{II} are Solid Phases I and II, respectively; L_t, L_m , and L_e are the latent heat at T_t, the heat of melting at T_m, and the heat of vapourisation at T_e , respectively; $C_P^{(\ell)}$ and $C_P^{(g)}$ are the C_P of the liquid and gas phases, respectively; and n is the number of moles.

Example 4.1

Calculate the molar enthalpy change of Mn, $\Delta\overline{H}_{Mn}$, between 1373 and 1573 K. The molar heat capacity (heat capacity per mole), $\overline{C}_P$ ($\mathrm{J \cdot mol^{-1} \cdot K^{-1}}$), latent heat at T_t , and heat of melting at temperature T_m per mole are as follows:

$$\overline{C}_P^{(\gamma)} = 25.23 + 14.90 \times 10^{-3}T - 1.854 \times 10^5 T^{-2} \quad (298\text{–}1410\,\mathrm{K})$$

$$\overline{C}_P^{(\delta)} = 46.44 \quad (1410\text{–}1517\,\mathrm{K})$$

$$\overline{C}_P^{(\ell)} = 46.02 \quad (1517\text{–}2324\,\mathrm{K})$$

$$\overline{L}_t(1410\,\mathrm{K}) = 1799\,\mathrm{J \cdot mol^{-1}}$$

$$\overline{L}_m(1517\,\mathrm{K}) = 14640\,\mathrm{J \cdot mol^{-1}}$$

Solution:

$$\begin{aligned}\Delta\overline{H}_{Mn} &= \int_{1373}^{1410} \overline{C}_P^{(\gamma)}\mathrm{d}T + \overline{L}_t + \int_{1410}^{1517} \overline{C}_P^{(\delta)}\mathrm{d}T + \overline{L}_m + \int_{1517}^{1573} \overline{C}_P^{(l)}\mathrm{d}T \\ &= \int_{1373}^{1410} \left(25.23 + 14.90 \times 10^{-3}T - 1.854 \times 10^5 T^{-2}\right)\mathrm{d}T \\ &\quad + 1799 + \int_{1410}^{1517} 46.44\mathrm{d}T + 14640 \\ &\quad + \int_{1517}^{1573} 40.02\mathrm{d}T \\ &= 1697 + 1799 + 4969 + 14640 + 2577 \\ &= 25682\,\mathrm{J \cdot mol^{-1}}\end{aligned}$$

4.6 Enthalpy of Formation (Heat of Formation)

When the compound A_xB_y is formed from its elements, A and B, the reaction can be described as follows:

$$xA + yB = A_xB_y \tag{4.27}$$

The enthalpy change of this kind of reaction is termed the *enthalpy of formation* (or *heat of formation*), $\Delta_f H$, and can be described as follows:

$$\Delta_f H_{A_xB_y} = H_{A_xB_y} - (xH_A + yH_B) \tag{4.28}$$

When all components are in their *standard states*, Eq. (4.28) can be rewritten as

$$\Delta_f H^\circ_{A_xB_y} = H^\circ_{A_xB_y} - \left(xH^\circ_A + yH^\circ_B\right) \tag{4.29}$$

$\Delta_f H^\circ_{A_xB_y}$ is the *standard enthalpy of formation* or *standard heat of formation* of A_xB_y , and is often stated alongside temperature. For example, the standard enthalpy of formation at 298.15 K is described as $\Delta_f H^\circ_{298}$. As is discussed in Sect. 4.2, the enthalpy of elements in most stable states at under 1 bar and 298.15 K (25 °C) is defined as zero, and thus both H°_A and H°_B in Eq. (4.29) equal zero at 298.15 K. Hence, $\Delta_f H^\circ_{298,A_xB_y} = H^\circ_{298,A_xB_y}$. Note that this relationship is true only at 298.15 K. The heat of formation of Element A in its reference state, i.e. $\Delta_f H^\circ_A$ of the reaction $A \rightarrow A$ (A at reference state) is clearly zero.

Example 4.2

Calculate the heat of formation of Al_2O_3 at 773 K (500 °C).

$$2Al(s) + \frac{3}{2}O_2(g) = Al_2O_3(s)$$

Solution:
According to thermodynamic data books,

$$\Delta H^\circ_{298} = -1675300\,\mathrm{J \cdot mol^{-1}}$$

$$\overline{C}^\circ_{P,Al_2O_3} = 114.8 + 12.80 \times 10^{-3}T - 35.44 \times 10^5 T^{-2}\,\mathrm{J \cdot mol^{-1} \cdot K^{-1}}\ (273 - 2000\,\mathrm{K})$$

$$\overline{C}^\circ_{P,Al} = 20.67 + 12.38 \times 10^{-3}T\,\mathrm{J \cdot mol^{-1} \cdot K^{-1}}\ (273 - 923\,\mathrm{K})$$

$$\overline{C}^{\circ}_{P,O_2} = 29.96 + 4.184 \times 10^{-3}T - 1.67 \times 10^{5}T^{-2}\,\text{J}\cdot\text{mol}^{-1}\cdot\text{K}^{-1}\ (273 - 2000\,\text{K})$$

where $\overline{C}^{\circ}_{P,i}$ is the molar heat capacity of component i in a standard state (P = 1 bar).
Therefore,

$$\Delta C^{\circ}_P = \overline{C}^{\circ}_{P,Al_2O_3} - \left(\frac{3}{2}\overline{C}^{\circ}_{P,O_2} + 2\overline{C}^{\circ}_{P,Al}\right)$$
$$= 28.52 - 18.24 \times 10^{-3}T - 32.93 \times 10^{5}T^{-2}$$

According to Eq. (4.22),

$$\Delta H^{\circ} = \Delta H_0^{\circ\prime} + 28.52\,T - \frac{18.24}{2} \times 10^{-3}T^2 + 32.93 \times 10^{5}T^{-1}$$

As $\Delta H^{\circ}_{298} = -1675300$ at T=298.15 K, the integral constant, $\Delta H_0^{\circ\prime} = -1694000$
Hence,

$$\Delta H^{\circ} = 28.52\,T - 9.12 \times 10^{-3}T^2 + 32.93 \times 10^{5}T^{-1} - 1694000\,\text{J}\cdot\text{mol}^{-1}$$

At T = 773 K,

$$\Delta H^{\circ}_{773} = -1673400\,\text{J}\cdot\text{mol}^{-1}$$

The ΔH°_{773} and ΔH°_{298} values show that, generally speaking, the temperature dependence of the heat of reaction is not significant.

Example 4.3

Compare the heat of reaction of the two SiO_2 (cristobalite) formation reactions at 1873 K (1600 °C).
(a)

$$Si(\ell) + O_2(g) = SiO_2(\text{cristobalite})$$

(b)

$$\underline{Si}(x_{Si} = 0.1) + O_2(g) = SiO_2(\text{cristobalite})$$

$\underline{Si}$ denotes Si dissolved in molten iron.
At 1873 K,

$$\overline{H}^{\circ}_{SiO_2(cri)} = -800600\,J \cdot mol^{-1}$$

$$\overline{H}^{\circ}_{Si(\ell)} = 91300\,J \cdot mol^{-1}$$

$$\overline{H}^{\circ}_{O_2(g)} = 53860\,J \cdot mol^{-1}$$

Solution:

(a) $\Delta H^{\circ}_{(a)} = \overline{H}^{\circ}_{SiO_2(cri)} - \overline{H}^{\circ}_{Si(\ell)} - \overline{H}^{\circ}_{O_2(g)}$
$= -800600 - 91300 - 53860$
$= -945800\,J \cdot mol^{-1}$

(b) $\Delta H_{(b)} = \Delta H^{\circ}_{(b)} + \Delta \overline{H}_{SiO_2(cri)} - \Delta \overline{H}_{Si(x_{Si}=0.1)} - \Delta \overline{H}^{\circ}_{O_2(g)}$
$= \overline{H}^{\circ}_{SiO_2(cri)} - \overline{H}^{\circ}_{Si(\ell)} - \overline{H}^{\circ}_{O_2(g)} + \Delta \overline{H}_{SiO_2(cri)} - \Delta \overline{H}_{Si(x_{Si}=0.1)}$
$- \Delta \overline{H}_{O_2(g)}$

where

$$\Delta \overline{H}_{SiO_2(cri)} = \overline{H}_{SiO_2(cri)} - \overline{H}^{\circ}_{SiO_2(cri)} = 0$$

$$\Delta \overline{H}_{O_2(g)} = \overline{H}_{O_2(g)} - \overline{H}^{\circ}_{O_2(g)} = 0$$

$$\Delta \overline{H}^{\circ}_{Si(x_{Si}=0.1)} = -125100\,J \cdot mol^{-1}$$

Hence,

$$\begin{aligned}\Delta H_{(b)} &= \overline{H}^{\circ}_{SiO_2(cri)} - \overline{H}^{\circ}_{Si(\ell)} - \overline{H}^{\circ}_{O_2(g)} - \Delta \overline{H}_{Si(x_{Si}=0.1)}\\ &= -800600 - 91300 - 53860 - (-125100)\\ &= -820660\,J \cdot mol^{-1}\end{aligned}$$

The difference between Reactions (a) and (b) is

$$\Delta H^{\circ}_{(a)} - \Delta H_{(b)} = \Delta \overline{H}_{Si(x_{Si}=0.1)} = -125140\,J \cdot mol^{-1}$$

4.7 Hess's Law

As is discussed above, enthalpy, H, is a state function. Therefore, the value of ΔH does not depend on the path of the change and is constant once the initial and final states of a system are specified. This fact is known as *Hess's law*.

In a general form, it can be expressed as follows:

$$\Delta_r H = \sum \Delta_f H(\text{products}) - \sum \Delta_f H(\text{reactants}) \tag{4.30}$$

Hess's law is useful in practice as the heat of reaction, which is difficult to measure, can be calculated by combining the heat of reaction of other reactions.

Thermodynamic data books list heat of reaction data for compounds. The heat of reaction of a given reaction can be calculated by combining the heat of reaction of other compounds.

As is discussed in Sect. 4.2, for the sake of convenience, the enthalpy of elements in most stable states at under 1 bar and 298.15 K is defined as zero. By using e.g. the relationship $\Delta_f H^\circ_{298,A_xB_y} = H^\circ_{298,A_xB_y}$, which is derived in Sect. 4.6 by defining $H^\circ_{298,i} \equiv 0$, and Hess's law, the enthalpy of formation of the reaction $CO_2(g)+C(s) = 2CO(g)$,

$$\Delta_f H^\circ_{(4.31)} = 2H^\circ_{CO} - \left(H^\circ_{CO_2} + H^\circ_{C}\right) \tag{4.31}$$

can be obtained as follows:

The enthalpy of formation of CO by $C + O = CO$ is

$$\Delta_f H^\circ_{CO} = H^\circ_{CO} - \left(H^\circ_{C} + H^\circ_{O}\right) = H^\circ_{CO} \tag{4.32}$$

and the enthalpy of formation of CO_2 by $C + 2O = CO_2$ is

$$\Delta_f H^\circ_{CO_2} = H^\circ_{CO_2} - \left(H^\circ_{C} + H^\circ_{O}\right) = H^\circ_{CO_2} \tag{4.33}$$

Hence, $\Delta_f H^\circ_{CO_2}$ of the reaction described in Eq. (4.31) is $2 \times$ Eq. (4.32) – Eq. (4.33)

$$\begin{aligned}\Delta_f H^\circ_{(4.31)} &= 2H^\circ_{CO} - \left(H^\circ_{CO_2} + H^\circ_{C}\right)\\ &= 2H^\circ_{CO} - H^\circ_{CO_2}\left(\text{because } H^\circ_{C} = 0 \text{ by definition}\right)\\ &= 2\Delta_f H^\circ_{CO} - \Delta_f H^\circ_{CO_2}\end{aligned} \tag{4.34}$$

Values for $\Delta_f H^\circ_{CO}$ and $\Delta_f H^\circ_{CO_2}$ are available in thermodynamic data books.

Example 4.4

Calculate the heat of reaction, $\Delta_r H^\circ_{(1)}$, of the following reaction:

$$MgO(s) + C(G) = Mg(g) + CO(g) \tag{1}$$

(s), (G), and (g) denote solid, graphite, and gas, respectively.
Use the following $\Delta_f H^\circ$ values for the calculation:

$$C(G) + \frac{1}{2}O_2(g) = CO(g)$$
$$\Delta_f H^\circ_{(2)} = -118000\,\text{J}\cdot\text{mol}^{-1} \tag{2}$$
$$Mg(g) + \frac{1}{2}O_2(g) = MgO(s)$$
$$\Delta_f H^\circ_{(3)} = -731100\,\text{J}\cdot\text{mol}^{-1} \tag{3}$$

Solution:
Reaction (1) = Reaction (2)–Reaction (3)
According to Hess's law,

$$\begin{aligned}\Delta_r H^\circ_{(1)} &= \Delta_f H^\circ_{(2)} - \Delta_f H^\circ_{(3)}\\ &= -118000 - (-731100)\\ &= 613100\,\text{J}\cdot\text{mol}^{-1}\end{aligned}$$

4.8 Summary

Enthalpy, H:

Enthalpy, H, is defined as

$$H \equiv U + PV \tag{4.1}$$

where U is the internal energy, P is the equilibrium pressure of a system, and V is the volume.

Heat capacity:

Heat capacity can be defined as follows:
Heat capacity at constant volume:

$$C_V = \left(\frac{\partial Q}{\partial T}\right)_V = \left(\frac{\partial U}{\partial T}\right)_V \tag{4.4}$$

Heat capacity at constant pressure:

$$C_P = \left(\frac{\partial Q}{\partial T}\right)_P = \left(\frac{\partial H}{\partial T}\right)_P \tag{4.5}$$

For 1 mol of an ideal gas, the relationship between C_V and C_P is

$$C_P = C_V + R \tag{4.13}$$

Temperature dependence of ΔH:

The temperature dependence of ΔH can be described by Kirchhoff's law.

$$\left(\frac{\partial \Delta H}{\partial T}\right)_P = \Delta C_P \tag{4.21}$$

ΔH values can be calculated using the following equation and C_P data obtained from a thermodynamic data book:

$$\Delta H = \int \Delta C_P \mathrm{d}T + \text{Constant} \tag{4.22}$$

Temperature dependence of enthalpy:

Enthalpy change against temperature can be calculated using the following equation:

$$\Delta \overline{H} = \frac{1}{n}\left(\int_{T_1}^{T_t} C_P^{(\mathrm{S_I})}\mathrm{d}T + L_t + \int_{T_t}^{T_m} C_P^{(\mathrm{S_{II}})}\mathrm{d}T + L_m + \int_{T_m}^{T_e} C_P^{(\ell)}\mathrm{d}T + L_e + \int_{T_e}^{T_2} C_P^{(g)}\mathrm{d}T\right) \tag{4.26}$$

Hess's law:

The value of ΔH depends only on the initial and final state of a system.

Chapter 5
The Second Law of Thermodynamics

The second law of thermodynamics is one of the most important concepts in thermodynamics, and a full understanding of it is required in order to understand the field as a whole. However, it is relatively complex and difficult to understand due to the complexity inherent in explaining it. Therefore, this book devotes considerable space to the second law of thermodynamics and one of the most important factors in its understanding; entropy.

5.1 Expressing the Second Law of Thermodynamics

The first law of thermodynamics describes the law of energy conservation. The second law of thermodynamics describes the direction of changes in energy and, in a broad sense, provides criteria that describe the direction of natural processes. The second law of thermodynamics can be described in many different ways, and the various textbooks on thermodynamics currently available provide at least 30 different descriptions. However, the scientific meaning (nature) of each of these different expressions is the same, and this fact makes it difficult to understand the second law of thermodynamics. There are several expressions of the second law of thermodynamics, but the following two are the most well-known:

1. Thomson's[1] principle (also known as Kelvin's principle)
   In a cycle of processes, it is impossible to transfer heat from a heat reservoir and convert it all into work without at the same time transferring a certain amount of heat from a hotter to a colder body.

[1]William Thomson (1st Baron Kelvin; 1824–1907) was a physicist and mathematician who contributed to many branches of classical physics, including thermodynamics, fluid dynamics, electromagnetics, and geophysics. In addition to his formulation of the second law of thermodynamics, he is also known for the Joule-Thomson effect, absolute temperature (Kelvin), and many other findings.

T. Matsushita and K. Mukai, *Chemical Thermodynamics in Materials Science*,
https://doi.org/10.1007/978-981-13-0405-7_5

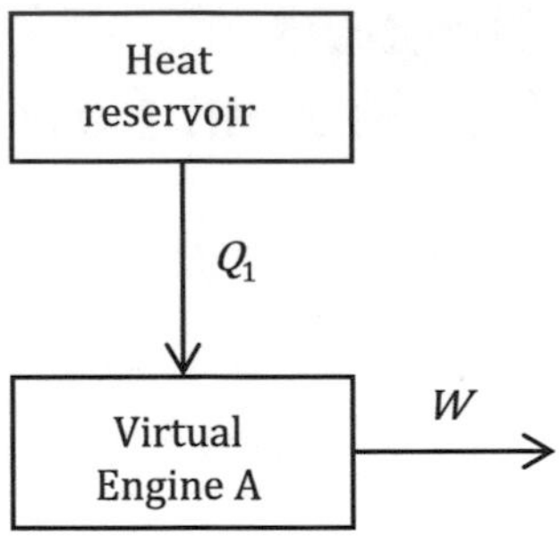

Fig. 5.1 An engine that violates Thomson's principle

2. Clausius's[2] principle
   It is impossible that, at the end of a cycle of changes, heat has been transferred from a colder to a hotter body without a certain amount of work having been converted into heat at the same time.

In the principles of Thomson and Clausius, the second law of thermodynamics is expressed very differently. However, both principles deal with the concept of an *irreversible process*—one wherein it is not possible to return to the initial state without any change in the system and surroundings. If it is possible, such a process is a *reversible process*. The details of reversible and irreversible processes are discussed in Sect. 5.2.

We can ascertain that these two principles are equivalent to each other using the contrapositive method.

Consider an engine that violates Thomson's principle; a virtual Engine A that can take heat Q_1 from a heat reservoir (the temperature of a heat reservoir is always constant despite heat being released to or absorbed from its surroundings) and do positive work, W (= Q_1), to its surroundings (Fig. 5.1).

Engine B in Fig. 5.2 can be used as a heat pump using the work, W, of virtual Engine A; heat is transferred from the lower-temperature heat reservoir to the higher-temperature heat reservoir (Fig. 5.2). This violates Clausius's principle, which states that heat (here $Q_1 + Q_2$) cannot be transferred from a colder body to a hotter one without at the same time converting a certain amount of work into heat.

Now, consider a virtual engine that violates Clausius's principle; a virtual Engine C that can transfer heat ($Q_3 > 0$) from a lower-temperature heat reservoir to a higher-temperature heat reservoir (Fig. 5.3). Engine D outputs work, W, (Q_4) using heats Q_3 and Q_4; the former (Q_3) is taken from the lower-temperature heat reservoir, while the latter (Q_4) is taken from the higher-temperature heat reservoir (total heat $= Q_3 + Q_4$; Fig. 5.3).

[2]Rudolf Clausius (1822–1888) was a physicist and mathematician who established the field of thermodynamics through the formulation of its first and second laws and introduction of the concept of entropy.

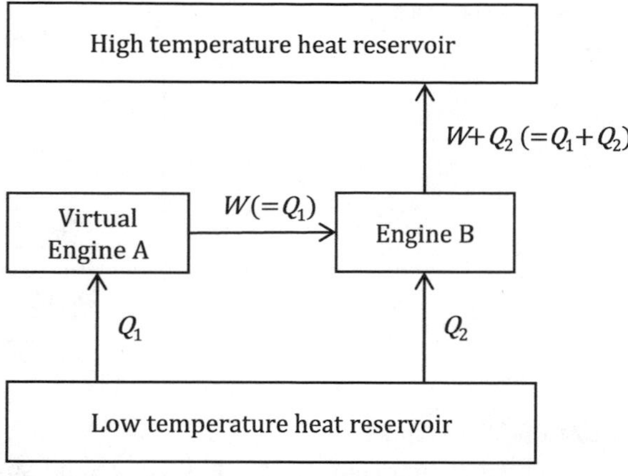

Fig. 5.2 A system that violates Clausius's principle

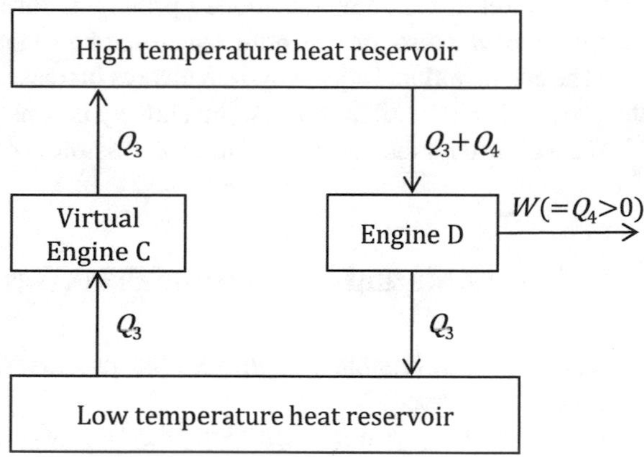

Fig. 5.3 A system that violates Clausius's principle and Thomson's principle

In this case, the net process is:

(1) Heat (Q_4) is transferred to Engine D.
(2) Engine D does the work, W (Q_4).

Here, Thomson's principle—which states that it is impossible to transfer heat (here Q_4) from a heat reservoir (here a higher-temperature heat reservoir) and convert it all into work, W (Q_4), without at the same time transferring a certain amount of heat from a hotter to a colder body—is violated.

Virtual engines A and C are *perpetual motion machines of the second kind*, and the second law of thermodynamics denies their existence. This fact is known as Ostwald's principle[3] and is also an expression of the second law of thermodynamics.

[3]Wilhelm Ostwald (1853–1932) was a chemist who received the Nobel Prize in Chemistry in 1909. He is regarded as one of the founders of the field of physical chemistry, together with Arrhenius and van't Hoff.

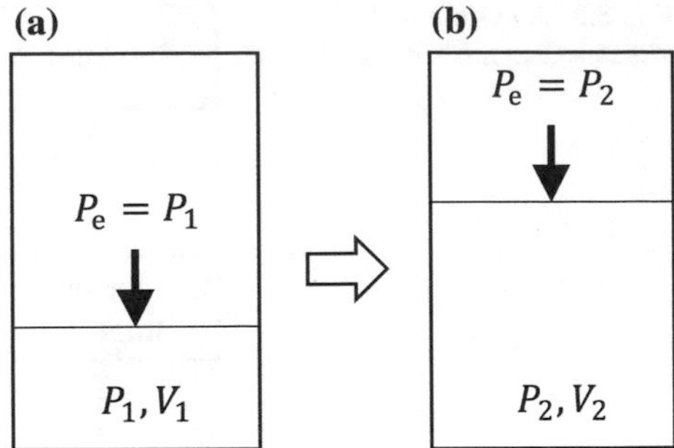

Fig. 5.4 A piston-cylinder system

Ostwald's principle:

"A perpetual motion of the second kind is impossible."

In addition to the above-mentioned principles for the second law of thermodynamics, consider also the following statement by Clausius:

"The entropy of an isolated system always increases in an irreversible process; if the process is a reversible process, the entropy is constant."

We will discuss the meaning of this expression later (Sect. 5.4).

5.2 Reversible and Irreversible Processes

In this section, reversible and irreversible processes, mentioned in the previous section, are discussed.

Assume that a frozen fish has been taken out of a freezer and absorbs heat from its surroundings, such that it reaches room temperature. To re-freeze the fish, we must put it in the freezer again. For the fish to become frozen in the freezer, its surroundings must provide the work to the system by using e.g. electrical energy. This is because the amount of energy that was used in the heating (thawing) process is not enough to freeze the fish again: It is impossible to revert to the initial state (i.e. a frozen fish) without extra work from the surroundings. The thawing process is an irreversible process and, in order to revert to the original state, an input of work from the surroundings is required.

Consider, on the other hand, a reversible process; one which allows state changes from State I to State II and back again to occur without any change during the process. In other words, the process is a path that goes through a series of infinitesimally small incremental steps, each of which is an equilibrium state. Thus, a reversible process is practically impossible, and is ideal and virtual.

We will now discuss the details of reversible and irreversible processes. Consider an ideal gas confined to a cylinder with a frictionless piston; the pressure of the ideal gas, P_1, and external pressure, P_e, are balanced (in an equilibrium state) (Fig. 5.4a).

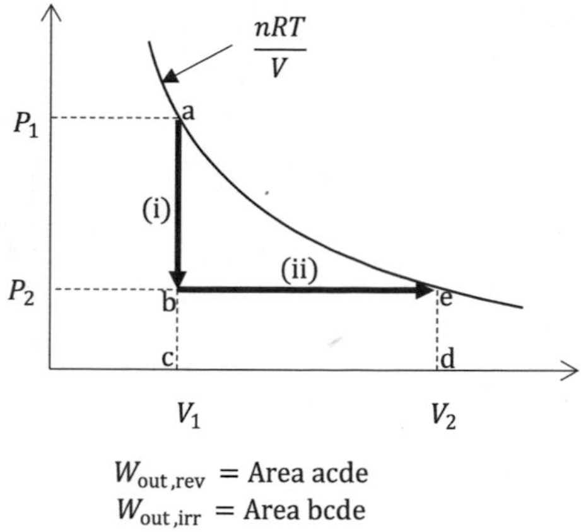

Fig. 5.5 Change in pressure and volume of a piston-cylinder system

Let us assume that the external pressure P_e decreases to P_2 over the course of a short period of time, and so the volume increases from V_1 to V_2 (Fig. 5.4b). In this case, the path of the change is (i) and (ii) in Fig. 5.5.

The work done by the system to its surroundings, $W_{out}(= -W_{in})$, is equal to the area bcde of Fig. 5.5. This is an irreversible process, and so is denoted by $W_{out,irr}$.

On the other hand, when the external pressure, P_e, is changed from P_1 to P_2 (and the volume from V_1 to V_2) over an infinite period of time, the process is reversible and the path of the change is the same as the curve nRT/V in Fig. 5.5. In this case, the work done by the system to the surroundings, $W_{out,rev}$, is equal to the area acde of Fig. 5.5. As can be seen from this example, we can extract the maximum work from a system when the process is reversible.

5.3 Carnot Engine (Carnot Cycle)

In this section, the 'engine' discussed in Sect. 5.1 is explored.

A *Carnot engine* is an idealised (i.e. reversible) engine which has the maximum possible efficiency. It is a reversible cycle, and extracts the maximum amount of work in the forward cycle; in the reverse cycle, it transfers heat from a lower-temperature heat reservoir to a higher-temperature heat reservoir with the minimum amount of work (an example is shown in Fig. 5.5).

The set of reversible processes is called the *Carnot cycle* and consists of four steps, after which it returns to the initial state.

(1) Isothermal expansion
(2) Adiabatic expansion
(3) Isothermal compression
(4) Adiabatic compression

Fig. 5.6 Isothermal expansion

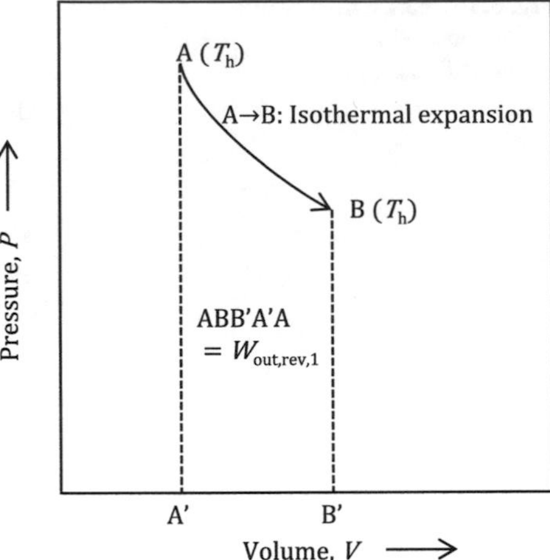

These changes of state occur quasi-statically over an infinite period of time, and the equilibrium state is always maintained.

Let us consider the Carnot cycle for an ideal gas (e.g. an ideal gas in a cylinder with a piston).

(1) Isothermal expansion

Isothermal expansion from Point A (Temperature: T_h) to Point B (Temperature: T_h) (Fig. 5.6).

The internal energy change, ΔU, of an ideal gas depends only on temperature (the definition of an ideal gas). Thus, $\Delta U = 0$ in the case of isothermal change ($\Delta T = 0$). Hence, the work done by the system to its surroundings for isothermal expansion, $W_{out,rev,1}$, is

$$W_{out,rev,1} = \int_{V_A}^{V_B} P \mathrm{d}V \tag{5.1}$$

$P = nRT/V$, thus

$$W_{out,rev,1} = nRT_h \int_{V_A}^{V_B} \frac{1}{V} \mathrm{d}V = nRT_h \ln \frac{V_B}{V_A} \tag{5.2}$$

The area ABB'A'A in Fig. 5.6 corresponds to the work, $W_{out,rev,1}$, which is equal to the heat, $Q_{in,rev,1}$, absorbed by the system during isothermal expansion.

(2) Adiabatic expansion

Adiabatic expansion from Point B (Temperature: T_h) to Point C (Temperature: T_ℓ) (Fig. 5.7).

According to the first law of thermodynamics,

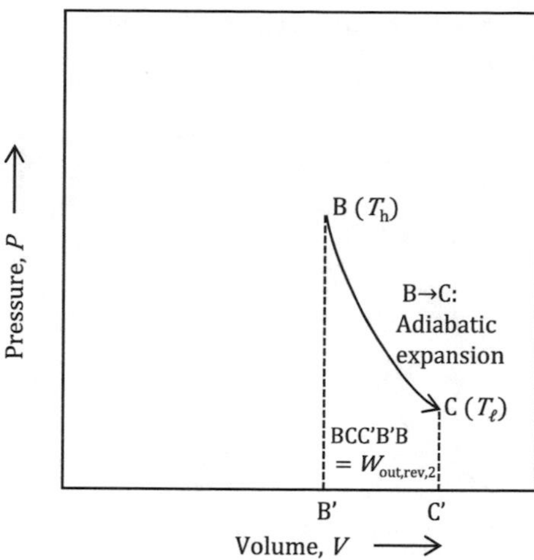

Fig. 5.7 Adiabatic expansion

$$\Delta U = Q_{in} - P\mathrm{d}V \tag{5.3}$$

In the case of adiabatic expansion, the heat absorbed by the system is zero ($Q_{in} = 0$). Thus,

$$\Delta U = -P\mathrm{d}V \tag{5.4}$$

and, as shown in Eq. (4.4),

$$\mathrm{d}U = C_V\mathrm{d}T \tag{5.5}$$

Hence, the work done by the system to its surroundings, $W_{out,rev,2}$, is

$$W_{out,rev,2} = \int_{V_B}^{V_C} P\mathrm{d}V = -\int_{T_h}^{T_\ell} C_V\mathrm{d}T = C_V(T_h - T_\ell) \tag{5.6}$$

The area BCC'B'B in Fig. 5.7 corresponds to work, $W_{out,rev,2}$.

(3) Isothermal compression

Isothermal compression from Point C (Temperature: T_ℓ) to Point D (Temperature: T_ℓ) (Fig. 5.8).

Unlike in the isothermal expansion case, with isothermal compression the work, $W_{in,rev,3}$, is done *to* the system and can be described as:

$$W_{in,rev,3} = nRT_\ell \ln\frac{V_D}{V_C} \tag{5.7}$$

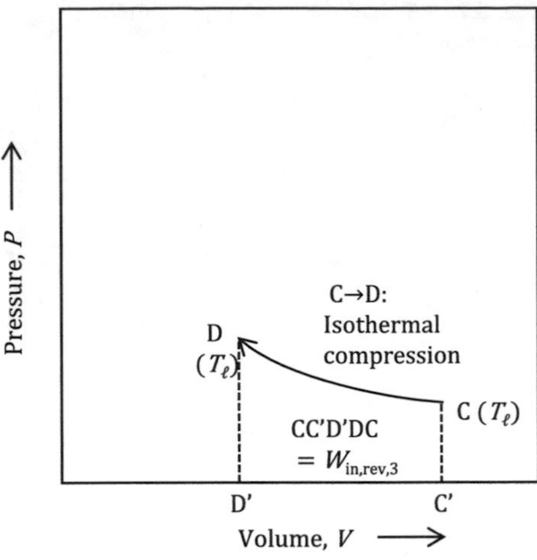

Fig. 5.8 Isothermal compression

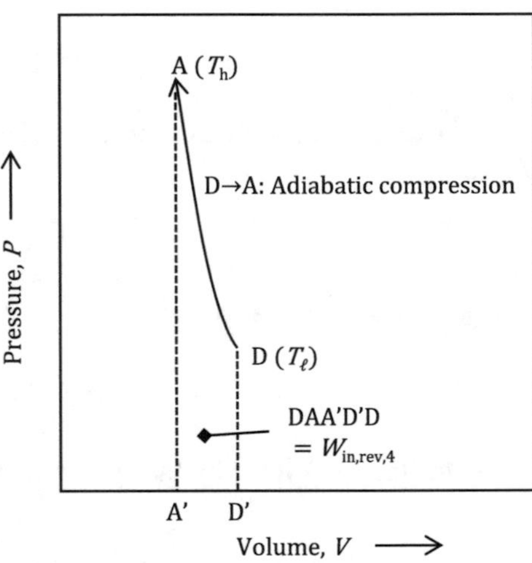

Fig. 5.9 Adiabatic compression

The area CC'D'DC in Fig. 5.8 corresponds to the work, $W_{\mathrm{in,rev,3}}$, which is equal to the heat, $Q_{\mathrm{out,rev,3}}$, released to the surroundings in order to maintain a constant temperature.

(4) Adiabatic compression

Adiabatic compression from Point D (Temperature: T_ℓ) to Point A (Temperature: T_h) (Fig. 5.9).

Unlike in the adiabatic expansion case, with adiabatic compression work, $W_{\mathrm{in,rev,4}}$, is done *to* the system and can be described as:

$$W_{\mathrm{in,rev,4}} = C_V(T_\mathrm{h} - T_\ell) \tag{5.8}$$

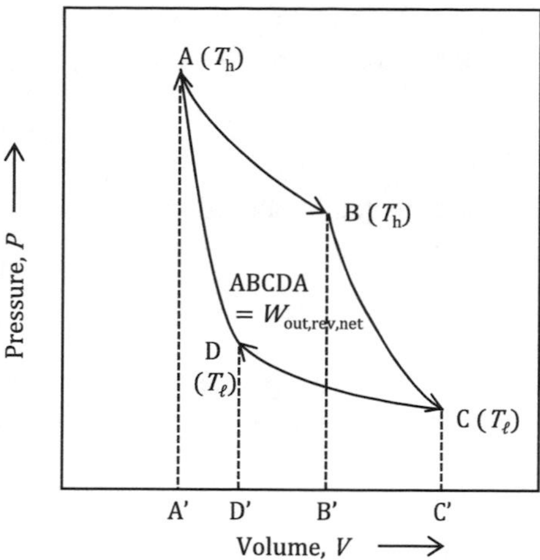

Fig. 5.10 Net work done by a system

The area DAA'D'D in Fig. 5.9 corresponds to work, $W_{in,rev,4}$.

Therefore, the net work done *by* the system, $W_{out,rev,net}$, after a cycle is

$$\begin{aligned} W_{out,rev,net} &= W_{out,rev,1} + W_{out,rev,2} - W_{in,rev,3} - W_{in,rev,4} \\ &= nRT_h \ln\left(\frac{V_B}{V_A}\right) - nRT_\ell \ln\left(\frac{V_C}{V_D}\right) \\ &= Q_{in,rev,1} - Q_{out,rev,3} \end{aligned} \tag{5.9}$$

The work done by the system ($W_{out,rev,1} + W_{out,rev,2}$) corresponds to the area ABCC'A'A, and the work done to the system ($W_{in,rev,3} + W_{in,rev,4}$) corresponds to the area CDAA'C'C in Fig. 5.10. Therefore, the area ABCDA in Fig. 5.10 corresponds to the net work done by the system, $W_{out,rev,net}$.

Based on the above discussion, it can be concluded that the Carnot cycle (engine) is a cycle that absorbs heat, $Q_{in,rev,1}$, from a heat reservoir with a higher temperature (T_h) and does work, $W_{out,rev,net}$, by releasing heat, $Q_{out,rev,3}$, to a heat reservoir with a lower temperature.

The ratio of the heat absorbed by the system to the work done by the system (engine) is termed *efficiency*, η, and is described by the following equation for the above-mentioned Carnot cycle:

$$\eta = \frac{W_{out,rev,net}}{Q_{in,rev,1}} = \frac{Q_{in,rev,1} - Q_{out,rev,3}}{Q_{in,rev,1}} \tag{5.10}$$

It can be also described as

$$\eta = \frac{T_h - T_\ell}{T_h} \tag{5.11}$$

The derivation of Eq. (5.11) is shown below.

Poisson's equation (*Poisson's law*) in thermodynamics,

$$TV^{\gamma-1} = \text{constant} \tag{5.12}$$

is valid for a quasi-static adiabatic change for an ideal gas.

$$\gamma \equiv \frac{C_P}{C_V} \tag{5.13}$$

where γ is the *heat capacity ratio.*

For the above-mentioned adiabatic expansion (Point B to C),

$$T_{\mathrm{h}} V_{\mathrm{B}}^{\gamma-1} = T_{\ell} V_{\mathrm{C}}^{\gamma-1} \tag{5.14}$$

and for the above-mentioned adiabatic compression (Point D to A),

$$T_{\ell} V_{\mathrm{D}}^{\gamma-1} = T_{\mathrm{h}} V_{\mathrm{A}}^{\gamma-1} \tag{5.15}$$

Thus,

$$\frac{T_{\mathrm{h}}}{T_{\ell}} = \frac{V_{\mathrm{C}}^{\gamma-1}}{V_{\mathrm{B}}^{\gamma-1}} = \frac{V_{\mathrm{D}}^{\gamma-1}}{V_{\mathrm{A}}^{\gamma-1}} \tag{5.16}$$

And so

$$\frac{V_{\mathrm{C}}}{V_{\mathrm{B}}} = \frac{V_{\mathrm{D}}}{V_{\mathrm{A}}} \tag{5.17}$$

Thus, the above-mentioned $W_{\mathrm{out,rev,net}}$ is

$$\begin{aligned} W_{\mathrm{out,rev,net}} &= W_{\mathrm{out,rev,1}} + W_{\mathrm{out,rev,2}} - W_{\mathrm{in,rev,3}} - W_{\mathrm{in,rev,4}} \\ &= Q_{\mathrm{in,rev,1}} - Q_{\mathrm{out,rev,3}} \\ &= nRT_{\mathrm{h}} \ln\left(\frac{V_{\mathrm{B}}}{V_{\mathrm{A}}}\right) - nRT_{\ell} \ln\left(\frac{V_{\mathrm{C}}}{V_{\mathrm{D}}}\right) \\ &= nR(T_{\mathrm{h}} - T_{\ell}) \ln\left(\frac{V_{\mathrm{B}}}{V_{\mathrm{A}}}\right) \end{aligned} \tag{5.18}$$

Hence,

$$\begin{aligned}\eta &= \frac{W_{\text{out,rev,net}}}{Q_{\text{in,rev},1}} = \frac{Q_{\text{in,rev},1} - Q_{\text{out,rev},3}}{Q_{\text{in,rev},1}} \\ &= \frac{nR(T_{\text{h}} - T_{\ell})\ln\left(\frac{V_{\text{B}}}{V_{\text{A}}}\right)}{nRT_{\text{h}}\ln\left(\frac{V_{\text{B}}}{V_{\text{A}}}\right)} \\ &= \frac{T_{\text{h}} - T_{\ell}}{T_{\text{h}}}\end{aligned} \tag{5.19}$$

(End of the derivation of Eq. (5.11))
As can be seen in the following equation,

$$\eta = \frac{T_{\text{h}} - T_{\ell}}{T_{\text{h}}} = 1 - \frac{T_{\ell}}{T_{\text{h}}} \tag{5.20}$$

efficiency, η, can be determined using only the temperature of the heat reservoirs with higher (T_{h}) and lower (T_{ℓ}) temperatures.

Based on the equation, efficiency is 1 (100 %) when the temperature of the heat reservoir with a higher temperature (T_{h}) is ∞. However, it is impossible to raise the temperature to ∞. $\eta = 1$ is also true when the temperature of the heat reservoir with a lower temperature (T_{ℓ}) is 0 K. This implies that all molecular motion is stopped completely, and all thermal motion can be converted to work. However, this is also impossible to achieve in practice.

From the equation, it is also possible to conclude that efficiency is zero and it is impossible to convert heat to work when the temperature of the two heat reservoirs is the same ($T_{\text{h}} = T_{\ell}$).

From the equation

$$\eta = 1 - \frac{Q_{\text{out,rev},3}}{Q_{\text{in,rev},1}} = 1 - \frac{T_{\ell}}{T_{\text{h}}} \tag{5.21}$$

the following relationship can be derived:

$$\frac{Q_{\text{out,rev},3}}{T_{\ell}} = \frac{Q_{\text{in,rev},1}}{T_{\text{h}}} \tag{5.22}$$

where Q_{rev}/T is termed *entropy*, S (discussed in Sect. 5.4 and Chap. 6). Based on this relationship, we can conclude that the *entropy change* of the system, ΔS, as regards the reversible change is zero, i.e.

$$\Delta S = \Delta S_1 + \Delta S_3 = -\frac{Q_{\text{in,rev},1}}{T_{\text{h}}} + \frac{Q_{\text{out,rev},3}}{T_{\ell}} = 0 \tag{5.23}$$

5.4 Mathematical Expression of the Second Law of Thermodynamics as Regards Entropy

In this section, a mathematical expression of the second law of thermodynamics will be derived.

As is shown in Eq. (3.1),

$$\Delta U = Q_{\text{in}} - W_{\text{out}} \tag{5.24}$$

where $W_{\text{out}} = P\mathrm{d}V$.

Therefore,

$$\Delta U = Q_{\text{in}} - P\mathrm{d}V \tag{5.25}$$

By dividing both sides of the equation by T, we obtain the following relationship:

$$\frac{\Delta U}{T} = \frac{Q_{\text{in}}}{T} - \frac{P\mathrm{d}V}{T} \tag{5.26}$$

In Eq. (5.26), internal energy, U, depends solely on the state of the system and is entirely independent of the path of the change. Hence, the change is equal to zero when the state is changed from State I to State II and back to State I. Now, consider a reversible process: The heat absorbed by the system can be denoted $Q_{\text{in,rev}}$. By using the contour integral $\oint$ (a type of integral used to denote a single integration over a loop), the above equation can be rewritten as

$$\oint \frac{\mathrm{d}U}{T} = \oint \frac{\mathrm{d}Q_{\text{in,rev}}}{T} - \oint \frac{P\mathrm{d}V}{T} = 0 \tag{5.27}$$

wherein the state changes from the initial one to another, then back to the initial state as a kind of cycle.

In the case of an ideal gas, $P/T = R/V$.

Therefore,

$$\frac{P\mathrm{d}V}{T} = \frac{R\mathrm{d}V}{V} \tag{5.28}$$

Hence,

$$\oint \frac{P\mathrm{d}V}{T} = \oint \frac{R\mathrm{d}V}{V} \tag{5.29}$$

The volume change after the cycle has been completed (from the initial state to another state, and back to the initial state via any path) is zero.

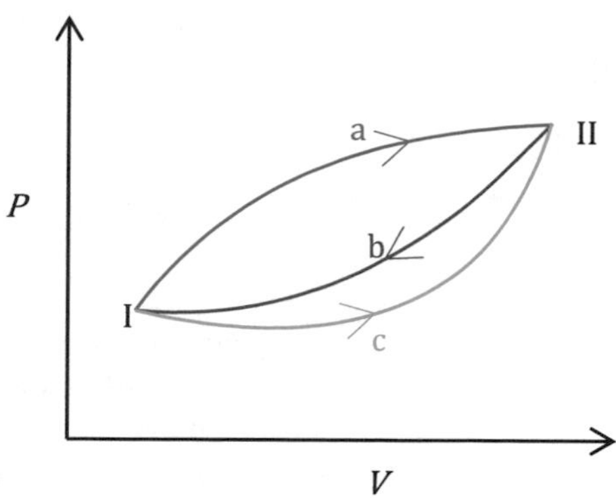

Fig. 5.11 A reversible process

Therefore,

$$\oint \frac{R\mathrm{d}V}{T} = R\mathrm{d}\ln V = 0 \tag{5.30}$$

Hence,

$$\oint \frac{\mathrm{d}Q_{\text{in,rev}}}{T} = 0 \tag{5.31}$$

As can be seen from these equations, the contour integral is zero. Therefore, $\frac{Q_{\text{in,rev}}}{T}$ is a state function, i.e. the value of $\frac{\mathrm{d}Q_{\text{in,rev}}}{T}$ does not depend on the path of the change. Hence, for the sake of convenience, we will define a new state function as follows:

$$\mathrm{d}S \equiv \frac{\mathrm{d}Q_{\text{in,rev}}}{T} \tag{5.32}$$

The new state function, S, is entropy.

In some books, the symbols δ, đ (d with a bar), or d' (d-prime) are used to denote an infinitesimal change in non-state functions such as work and heat. Therefore, Eq. (5.32) can be written as $\mathrm{d}S \equiv \frac{\delta Q_{\text{in,rev}}}{T}$. However, in this book, for the sake of simplicity, d is used for an infinitesimal change in non-state functions as well.

Now, we will consider a simple example of entropy change. When the state of a system changes from State I to State II via Path a and from State II to State I via Path b (Fig. 5.11) and it is assumed that the change is reversible, the entropy change can be described as follows:

$$\oint_{\text{I}\to\text{a}\to\text{II}\to\text{b}\to\text{I}} \frac{\mathrm{d}Q_{\text{in,rev}}}{T} = \int_{\text{I}\to\text{a}\to\text{II}} \frac{\mathrm{d}Q_{\text{in,rev}}}{T} + \int_{\text{II}\to\text{b}\to\text{I}} \frac{\mathrm{d}Q_{\text{in,rev}}}{T} = 0 \tag{5.33}$$

In a similar manner, when the change in state occurs as I → c → II → b → I,

$$\int_{\mathrm{I}\to\mathrm{c}\to\mathrm{II}} \frac{\mathrm{d}Q_{\mathrm{in,rev}}}{T} + \int_{\mathrm{II}\to\mathrm{b}\to\mathrm{I}} \frac{\mathrm{d}Q_{\mathrm{in,rev}}}{T} = 0 \tag{5.34}$$

Hence,

$$\int_{\mathrm{I}\to\mathrm{a}\to\mathrm{II}} \frac{\mathrm{d}Q_{\mathrm{in,rev}}}{T} = \int_{\mathrm{I}\to\mathrm{c}\to\mathrm{II}} \frac{\mathrm{d}Q_{\mathrm{in,rev}}}{T} \tag{5.35}$$

This equation implies that the entropy change resulting from the change in state ($\mathrm{I} \to \mathrm{II}$) does not depend on the path of the change.

Hence, it can be concluded that the value of $\frac{\mathrm{d}Q_{\mathrm{in,rev}}}{T}$ depends only on the state of the system and not on the path of the change. The integrated value

$$\int_{\mathrm{I}}^{\mathrm{II}} \frac{\mathrm{d}Q_{\mathrm{in,rev}}}{T} \tag{5.36}$$

is the entropy of State II relative to that of State I ($S_{\mathrm{I}\to\mathrm{II}}$), or the difference in entropy between State I and State II (ΔS).

Having extensively explored reversible processes, let us now consider the entropy change for irreversible processes.

As explained in Sect. 5.3, one can extract the maximum work, $W_{\mathrm{out,rev,net}}$, from a system when the change is reversible. Hence, the work one can extract in the case of an irreversible process, $W_{\mathrm{out,irr,net}}$, is always smaller than $W_{\mathrm{out,rev,net}}$.

$$W_{\mathrm{out,rev,net}} > W_{\mathrm{out,irr,net}} \tag{5.37}$$

Therefore,

$$Q_{\mathrm{in,rev}} > Q_{\mathrm{in,irr}} \tag{5.38}$$

By dividing both sides of the equation by T, we obtain

$$\frac{Q_{\mathrm{in,rev}}}{T} > \frac{Q_{\mathrm{in,irr}}}{T} \tag{5.39}$$

As is already defined,

$$\mathrm{d}S \equiv \frac{\mathrm{d}Q_{\mathrm{in,rev}}}{T} \tag{5.40}$$

Therefore, in the case of an irreversible process,

$$\mathrm{d}S = \frac{\mathrm{d}Q_{\mathrm{in,rev}}}{T} > \frac{\mathrm{d}Q_{\mathrm{in,irr}}}{T} \tag{5.41}$$

This equation clearly delineates reversible and irreversible processes. Moreover, based on this discussion, the following statements can be made regarding entropy:

1. For a reversible process,

$$\oint \frac{\mathrm{d}Q_{\mathrm{in,rev}}}{T} = \oint \mathrm{d}S = 0 \tag{5.42}$$

is true. Conversely, one might say that if the above equation is satisfied, the process is reversible.
2. Entropy, S, is a state function as $\oint \mathrm{d}S = 0$.
3. Entropy change can be calculated only when they occur during a reversible process.
4. For a reversible process,

$$\mathrm{d}S = \frac{\mathrm{d}Q_{\mathrm{in,rev}}}{T} \tag{5.43}$$

For an irreversible process,

$$\mathrm{d}S > \frac{\mathrm{d}Q_{\mathrm{in,irr}}}{T} \tag{5.44}$$

Using these equations, the second law of thermodynamics can be expressed as follows:

"The entropy of an isolated system always increases in an irreversible process; if the process is a reversible process, the entropy is constant." (Sect. 5.1).

5. For a reversible process,

$$\oint \frac{\mathrm{d}Q_{\mathrm{in,rev}}}{T} = 0 \tag{5.45}$$

For an irreversible process,

$$\oint \frac{\mathrm{d}Q_{\mathrm{in,irr}}}{T} < 0 \tag{5.46}$$

The *Clausius inequality* states that

$$\oint \frac{\mathrm{d}Q_{\mathrm{in}}}{T} \leq 0 \tag{5.47}$$

6. In reality, all changes are irreversible. Therefore, only an equilibrium state can be regarded as a reversible process (state). Hence,

$$\oint \mathrm{d}S = 0 \tag{5.48}$$

is a condition for equilibrium, and is discussed in Chap. 7.

5.5 Summary

The second law of thermodynamics:

The second law of thermodynamics can be described in multiple ways, but the meaning of each statement is identical. Thus:

"The entropy of an isolated system always increases in an irreversible process; if the process is a reversible process, the entropy is constant."

Reversible and irreversible processes:

A reversible process is one that causes no change during the change of state from State I to State II, and State II to State I. In other words, the process is a path that goes through a succession of equilibrium states. If a process does not satisfy this condition, it is irreversible.

Carnot cycle:

The Carnot cycle consists of 4 processes:

(1) Isothermal expansion
(2) Adiabatic expansion
(3) Isothermal compression
(4) Adiabatic compression

Efficiency can be described as

$$\eta = \frac{T_{\mathrm{h}} - T_{\ell}}{T_{\mathrm{h}}} \tag{5.11}$$

Entropy, S:

Entropy can be defined as

$$\mathrm{d}S \equiv \frac{\mathrm{d}Q_{\mathrm{in,rev}}}{T} \tag{5.32}$$

For a reversible process,

$$\mathrm{d}S = \frac{\mathrm{d}Q_{\mathrm{in,rev}}}{T} \tag{5.43}$$

For an irreversible process,

$$\mathrm{d}S > \frac{\mathrm{d}Q_{\mathrm{in,irr}}}{T} \tag{5.44}$$

These equations are mathematical expressions of the second law of thermodynamics.

Chapter 6
Entropy, S

The concept of entropy was introduced by Clausius through his investigation of the Carnot cycle, which was discussed in the previous chapter. Subsequently, it has been applied to chemical thermodynamics. The nature of entropy is the same in both general thermodynamics (the Carnot cycle) and chemical thermodynamics, but the concept is applied differently in the latter. This may hinder a full understanding of entropy in relation to chemical thermodynamics, and so is discussed from different perspectives in this book. In this chapter, entropy is discussed from microscopic and macroscopic perspectives in order to deepen our understanding of the concept.

6.1 The Microscopic Perspective on Entropy

The meaning of entropy, S, which is discussed in the previous chapter, will now be considered from a microscopicperspective. From a macroscopic perspective, entropy, S, can be described by the number of possible arrangements (microscopic states) of the system, $\mathcal{W}$.

$$S = k \ln \mathcal{W} \tag{6.1}$$

where k is the Boltzmann constant. This equation is called *Boltzmann's relation*. The microscopic state of a system, which consists of a great number of atoms and molecules distributed in a space, is defined by the arrangements of the atoms/molecules within the space, the vibration of the atoms/molecules, the state of rotation, and the positions of electrons. If the degree of freedom of the arrangements of the atoms/molecules, the state of vibration, the state of rotation, or the positions of the electrons is increased, $\mathcal{W}$ increases and so S also increases. Hence, entropy, S, can be regarded as the degree of randomness of a system.

We will consider a simple virtual experiment to understand the meaning of $\mathcal{W}$: A container is split into two parts by a partition (Fig. 6.1). Assume that each partition contains two gas molecules; two of A on the left side of the container, and two of

T. Matsushita and K. Mukai, *Chemical Thermodynamics in Materials Science*,
https://doi.org/10.1007/978-981-13-0405-7_6

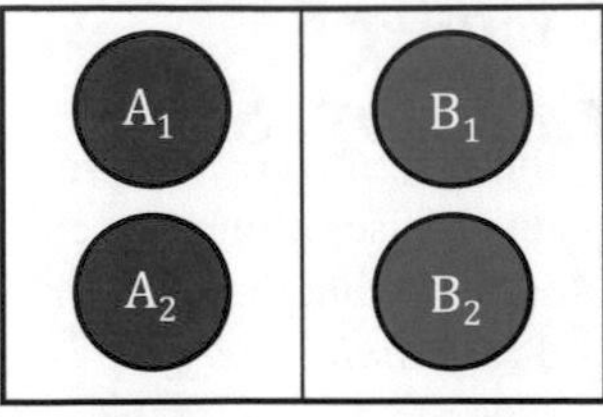

Fig. 6.1 Separated molecules

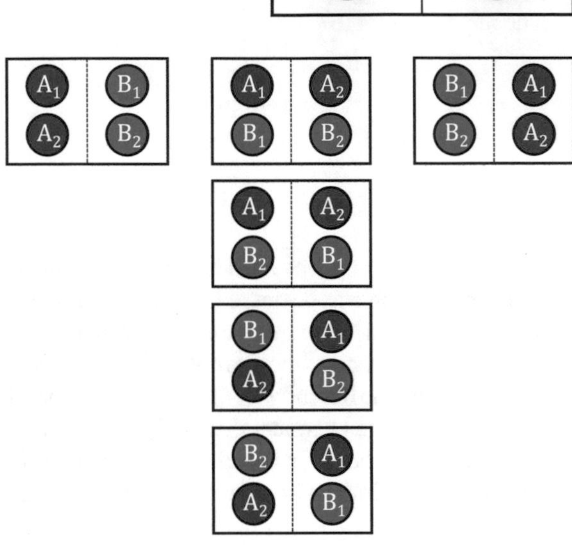

Fig. 6.2 Possible states

B on the right side. How are these molecules distributed in the container when the partition is removed?

As shown in Fig. 6.2, the number of possible states is

$$\frac{(N_A + N_B)!}{N_A!\,N_B!} = \frac{4!}{2!\,2!} = 6 \tag{6.2}$$

Which is to say that there are six configurations. This number corresponds to the number of possible arrangements (microscopic states) of the system—$\mathcal{W}$ in Boltzmann's relation.

There are four configurations wherein four molecules (two of A and two of B) mix uniformly in the container, with one Molecule A and one Molecule B on each side, and the probability of one of these configurations occurring is 4/6 = 2/3. On the other hand, the probability is 1/6 that the two Molecules A will congregate at the left side of the container, and the two Molecules B at the right side. Further, the two Molecules A may be at the right side, and the two Molecules B at the left side, and this event also has a probability of 1/6. In this example, the total number of molecules is extremely low (4), and so the probability of each of the configurations occurring is relatively high (1/6). However, when the number of molecules is high enough (for example, 1 mol, 6.02×10^{23} molecules), the probability of all Molecules A congregating at the left side of the container and all Molecules B at the right side is lower, approaching zero—meaning that, practically speaking, it is extremely unlikely to occur.

6.2 Calculation of Entropy

Although entropy as defined by the second law of thermodynamics may have little practical use, it must be considered in order to derive the important factors that are used to predict the equilibrium state of a thermodynamic system—*Gibbs energy*, G, and *Helmholtz energy*, F—e.g. to calculate $G = H - TS$.

The methods used to calculate entropy are described using two case studies below.

6.2.1 *Entropy Change During the Isothermal Expansion of Gas*

A non-ideal gas can be regarded as an ideal gas at relatively low pressures.

$$\mathrm{d}S = \frac{\mathrm{d}Q_{\mathrm{in,rev}}}{T} \tag{6.3}$$

Equation (6.3) states that $\mathrm{d}Q_{\mathrm{in,rev}}$ is the heat absorbed by a system during a reversible process.

By considering the reversible expansion of n mole of an ideal gas, the work done by a system, $\mathrm{d}W_{\mathrm{out}}$, during volume expansion, $\mathrm{d}V$, can be expressed as

$$\mathrm{d}W_{\mathrm{out}} = P\mathrm{d}V = nRT\frac{\mathrm{d}V}{V} \tag{6.4}$$

The internal energy of an ideal gas is a function only of temperature. Therefore,

$$\mathrm{d}U = \mathrm{d}Q_{\mathrm{in,rev}} - \mathrm{d}W_{\mathrm{out}} = 0 \tag{6.5}$$

Thus,

$$\mathrm{d}S = \frac{\mathrm{d}Q_{\mathrm{in,rev}}}{T} = \frac{\mathrm{d}W_{\mathrm{out}}}{T} = nR\frac{\mathrm{d}V}{V} \tag{6.6}$$

The entropy change, ΔS, during an expansion of volume from V_1 to V_2 is

$$\Delta S = \int_{V_1}^{V_2} \mathrm{d}S = nR\int_{V_1}^{V_2} \frac{\mathrm{d}V}{V} = nR \ln\left(\frac{V_2}{V_1}\right) \tag{6.7}$$

6.2.2 Temperature Dependence of Entropy, S

Under a constant pressure, $\mathrm{d}Q_{\mathrm{in,rev}}$ is described as follows:

$$\mathrm{d}Q_{\mathrm{in,rev}} = C_P \mathrm{d}T \tag{6.8}$$

Thus,

$$\mathrm{d}S = \frac{C_P \mathrm{d}T}{T} \tag{6.9}$$

Hence, the molar entropy change, $\Delta\overline{S}$, resulting from the change in temperature from T_1 to T_2 is

$$\begin{aligned}\Delta\overline{S} &= \overline{S}(T_2, P) - \overline{S}(T_1, P) \\ &= \frac{1}{n}\int_{T_1}^{T_2} \frac{C_P}{T}\mathrm{d}T = \int_{T_1}^{T_2} \frac{\overline{C}_P}{T}\mathrm{d}T\end{aligned} \tag{6.10}$$

Note that C_P is a function of temperature, and should not be placed outside the integration.

As with enthalpy (Eq. (4.26)), by substituting C_P into Eq. (6.10) and performing an integration, $\Delta\overline{S}$ can be calculated. If phase transformation, melting, and/or evaporation occurs during the temperature change between T_1 and T_2, the latent heat that accompanies these changes must be taken into account. Namely, the following terms must be added:

$$\text{Entropy of transformation}: \Delta S_{\mathrm{t}} = \frac{L_{\mathrm{t}}}{T_{\mathrm{t}}} \tag{6.11}$$

$$\text{Entropy of melting}: \Delta S_{\mathrm{m}} = \frac{L_{\mathrm{m}}}{T_{\mathrm{m}}} \tag{6.12}$$

$$\text{Entropy of vapourisation}: \Delta S_{\mathrm{e}} = \frac{L_{\mathrm{e}}}{T_{\mathrm{e}}} \tag{6.13}$$

In most cases these reactions occur above room temperature, and so it is convenient for the value of entropy at 298.15 K, S°_{298} . S°_{298} is termed *standard entropy*, the values for which can be found in thermodynamic data books. According to *the third law of thermodynamics*, "the entropy of a perfect crystal at 0 K, S_0, is zero". Hence, the absolute value of entropy can be calculated.

At 298.15 K, standard entropy, S°_{298}, is,

$$\begin{aligned}S^\circ_{298} &= S^\circ_{298} - S^\circ_0 = \int_0^{298} \frac{C^\circ_P}{T}\mathrm{d}T \\ &= \int_0^{298} C^\circ_P \mathrm{d}\ln T\end{aligned} \tag{6.14}$$

Note: According to the mathematical formula, $\frac{\mathrm{d}\ln T}{\mathrm{d}T} = 1/T$, $\frac{\mathrm{d}T}{T} = \mathrm{d}\ln T$.

The integration of the above equation can be performed using graphical integration by extrapolating the heat capacity value obtained from the thermodynamic data book towards 0 K.

Standard entropy at temperature T, S°_T, can be described as follows:

$$S^\circ_T = S^\circ_{298} + \int_{298}^{T} \frac{C^\circ_P}{T} \mathrm{d}T \tag{6.15}$$

If phase transformation(s) occur(s) between 298.15 and T K, the terms $\Delta S^\circ_\mathrm{t}$, $\Delta S^\circ_\mathrm{m}$, and $\Delta S^\circ_\mathrm{e}$ should be added accordingly.

In general, entropy change can be described as follows:

$$\Delta\overline{S} = \frac{1}{n}\left[\int_{T_1}^{T_\mathrm{t}} \frac{C_P^{(\mathrm{S_I})}}{T}\mathrm{d}T + \frac{L_\mathrm{t}}{T_\mathrm{t}} + \int_{T_\mathrm{t}}^{T_\mathrm{m}} \frac{C_P^{(\mathrm{S_{II}})}}{T}\mathrm{d}T + \frac{L_\mathrm{m}}{T_\mathrm{m}} + \int_{T_\mathrm{m}}^{T_\mathrm{e}} \frac{C_P^{(\ell)}}{T}\mathrm{d}T + \frac{L_\mathrm{e}}{T_\mathrm{e}} + \int_{T_\mathrm{e}}^{T_2} \frac{C_P^{(\mathrm{g})}}{T}\mathrm{d}T\right] \tag{6.16}$$

where $\mathrm{S_I}$ and $\mathrm{S_{II}}$ are Solid Phases I and II, respectively; L_t, L_m, and L_e are the latent heat at T_t, the heat of melting at T_m, and the heat of vapourisation at T_e, respectively; $C_P^{(\ell)}$ and $C_P^{(\mathrm{g})}$ are the C_P of the liquid phase and the gas phase, respectively; and n is the number of moles.

Example 6.1

Calculate the molar entropy change, $\Delta\overline{S}_\mathrm{Mn}$, when 1 mol of Mn is heated from 1373 to 1573 K.

Solution:

According to thermodynamic data books,

The molar heat capacity (heat capacity per mole) of Mn is:

$$\overline{C}_P^{(\gamma)} = 25.23 + 14.90\times10^{-3}T - 1.854\times10^{5}T^{-2}\,\mathrm{J\cdot mol^{-1}\cdot K^{-1}}(298-1410\ \mathrm{K})$$

$$\overline{C}_P^{(\delta)} = 46.44\,\mathrm{J\cdot mol^{-1}\cdot K^{-1}}\ (1410-1517\ \mathrm{K})$$

$$\overline{C}_P^{(\ell)} = 46.02\,\mathrm{J\cdot mol^{-1}\cdot K^{-1}}\ (1517-2324\ \mathrm{K})$$

Latent heat per mole is:

$$\overline{L}_\mathrm{t}(1410\ \mathrm{K}) = 1799\,\mathrm{J\cdot mol^{-1}}$$

$$\overline{L}_{\mathrm{m}}(1517\,\mathrm{K}) = 14640\,\mathrm{J}\cdot\mathrm{mol}^{-1}$$

Therefore,

$$\begin{aligned}\Delta\overline{S}_{\mathrm{Mn}} &= \int_{1373}^{1410}\frac{\overline{C}_P^{(\gamma)}}{T}\mathrm{d}T + \Delta\overline{S}_{\mathrm{t}} + \int_{1410}^{1517}\frac{\overline{C}_P^{(\delta)}}{T}\mathrm{d}T + \Delta\overline{S}_{\mathrm{m}} + \int_{1517}^{1573}\frac{\overline{C}_{\mathrm{P}}^{(\ell)}}{T}\mathrm{d}T \\ &= \int_{1373}^{1410}\frac{25.23 + 14.90\times10^{-3}T - 1.854\times10^{5}T^{-2}}{T}\mathrm{d}T \\ &\quad + \frac{1799}{1410} + \int_{1410}^{1517}\frac{46.44}{T}\mathrm{d}T + \frac{14640}{1517} + \int_{1517}^{1573}\frac{46.02}{T}\mathrm{d}T \\ &= 17.2\,\mathrm{J}\cdot\mathrm{mol}^{-1}\cdot\mathrm{K}^{-1}\end{aligned}$$

6.3 Macroscopic and Microscopic Perspectives on Entropy

We know that entropy is described as $\mathrm{d}S = \frac{\mathrm{d}Q_{\mathrm{in,rev}}}{T}$ from a macroscopic perspective, and $S = k\ln\mathcal{W}$ from a microscopic one. These two descriptions, however, seem completely different, so are they really equivalent? In this section, this issue is explored through the calculation of the entropy of mixing.

Consider an insulating container that is split into two parts by a partition. Assume that there are n_1 moles of Ideal Gas 1 on the left side of the container (of volume V_1) and n_2 moles of Ideal Gas 2 on the right side of the container (of volume V_2). The temperature is T and the pressure P for both sides (State I).

After the partition is removed, the two ideal gases will mix homogeneously and reach an equilibrium state (State II). In the following discussion, the change in entropy as a result of mixing, i.e. *entropy of mixing*, when the volume, V, becomes $V_1 + V_2$ at temperature T and pressure P, and the number of particles $n = n_1 + n_2$, is calculated.

6.3.1 Macroscopic Perspective

It is not possible to calculate the entropy of the above-described process due to the fact that it is irreversible. However, by finding a path between State I and State II that is quasi-static, it is possible to calculate entropy (state function), as the change is defined by the initial and final states. A virtual experiment involving a semipermeable membrane is presented below.

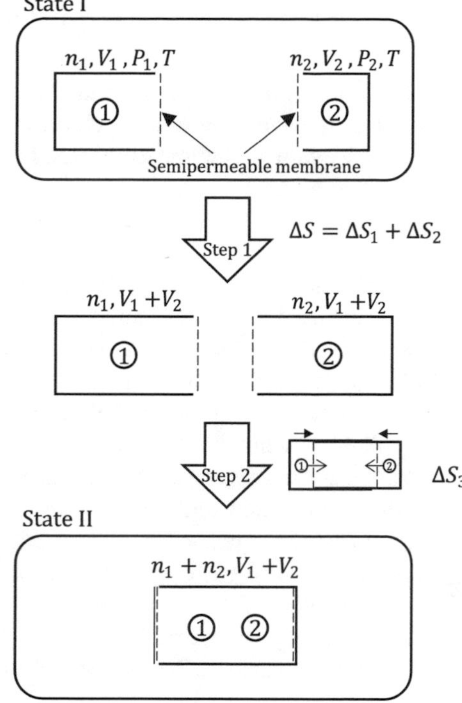

Fig. 6.3 Entropy of mixing (virtual experiment)

Firstly, we will separate Ideal Gas 1 and Ideal Gas 2 at State I into two different containers, and then we will increase the volume of both gases to $V_1 + V_2$ with adiabatic free expansion (Fig. 6.3, Step 1).

Adiabatic free expansion is an irreversible process. However, as we are working with an ideal gas, there is no temperature change during the process. Therefore, we can replace it with an isothermal expansion process, which is reversible, in order to calculate the entropy change. As is discussed above (Eq. (6.7)), the entropy change as a result of the change in volume, from V_1 to V_2, with isothermal expansion is

$$\Delta S = \int_{V_1}^{V_2} \mathrm{d}S = nR \int_{V_1}^{V_2} \frac{\mathrm{d}V}{V} = nR \ln\left(\frac{V_2}{V_1}\right) \tag{6.17}$$

Hence, in the case of isothermal expansion from V_1 to $V_1 + V_2$,

$$\Delta S_1 = \int_{V_1}^{V_1+V_2} \mathrm{d}S = n_1 R \int_{V_1}^{V_1+V_2} \frac{\mathrm{d}V}{V} = n_1 R \ln\left(\frac{V_1 + V_2}{V_1}\right) \tag{6.18}$$

In the case of an ideal gas, $V_1 = n_1(V_1 + V_2)$ and $V_2 = n_2(V_1 + V_2)$. Therefore,

$$\Delta S_1 = n_1 R \ln\left(\frac{n_1 + n_2}{n_1}\right) = -n_1 R \ln\left(\frac{n_1}{n_1 + n_2}\right) = -n_1 R \ln x_1 \tag{6.19}$$

where x_1 is the mole fraction of Ideal Gas 1.

In a similar manner, the entropy change, ΔS_2, when Ideal Gas 2 is expanded from V_1 to $V_1 + V_2$ with isothermal expansion is

$$\Delta S_2 = -n_2 R \ln x_2 \tag{6.20}$$

Imagine a semipermeable membrane in the container of Ideal Gas 1 which only allows Ideal Gas 2 to pass, and another such membrane in the container of Ideal Gas 2 which only allows Ideal Gas 1 to pass. By superimposing these two containers, Ideal Gas 1 will move into the container of Ideal Gas 2 through the semipermeable membrane, and vice versa. After the process, the mixture of Ideal Gas 1 and Ideal Gas 2, the volume of which is $V_1 + V_2$, is obtained (Fig. 6.3, Step 2). By performing this process quasi-statically, we are able to calculate entropy change. No work is involved in the quasi-static adiabatic superimposition process; there is no heat exchange, no partial pressure change for Ideal Gas 1 and Ideal Gas 2, no change in volume (we are considering ideal gases, and thus the *ideal gas law* and *Dalton's law* are true. Hence, the volume of the gas mixture is $V_1 + V_2$), and no change in internal energy. Hence, the entropy change of this process, ΔS_3, is zero.

The entropy change, ΔS, between State I and State II is

$$\begin{aligned}\Delta S = \Delta S_1 + \Delta S_2 &= -n_1 R \ln x_1 - n_2 R \ln x_2 \\ &= -R(n_1 \ln x_1 + n_2 \ln x_2)\end{aligned} \tag{6.21}$$

6.3.2 *Microscopic Perspective*

In the above calculations, the definition of entropy from a macroscopic perspective, $\mathrm{d}S \equiv \frac{\mathrm{d}Q_{\mathrm{in,rev}}}{T}$, was used to calculate the entropy change. Now, we will undertake the same process using Boltzmann's relation, $S = k \ln \mathcal{W}$.

Consider N_1 molecules of Ideal Gas 1 (of volume V_1) and N_2 molecules of Ideal Gas 2 (of volume V_2) in separate containers. We will calculate the entropy change when these gases are mixed and become $N_1 + N_2$ molecules, the volume of which is $V_1 + V_2$.

The number of possible arrangements (microscopic states) of the system is 1 before the mixing, i.e. $\mathcal{W}_\mathrm{I} = 1$. After the mixing, it becomes

$$\mathcal{W}_\mathrm{II} = \frac{(N_1 + N_2)!}{N_1!\, N_2!} \tag{6.22}$$

Hence, according to Boltzmann's relation,

$$\Delta S = k \ln \mathcal{W}_{\mathrm{II}} - k \ln \mathcal{W}_{\mathrm{I}} = k \ln \mathcal{W}_{\mathrm{II}} = k \ln \frac{(N_1 + N_2)!}{N_1!\, N_2!} \tag{6.23}$$

where k is the Boltzmann constant.

When N is sufficiently large, In $N! = N \ln N - N$; this is known as *Stirling's approximation*.

Therefore,

$$\begin{aligned} \Delta S &= k \ln \frac{(N_1 + N_2)!}{N_1!\, N_2!} \\ &= k\{[(N_1 + N_2)\ln(N_1 + N_2) - (N_1 + N_2)] \\ &\quad -[N_1 \ln N_1 - N_1 + N_2 \ln N_2 - N_2]\} \\ &= k(N_1 + N_2)\left[\ln(N_1 + N_2) - \frac{N_1 \ln N_1}{(N_1 + N_2)} - \frac{N_2 \ln N_2}{(N_1 + N_2)}\right] \end{aligned} \tag{6.24}$$

$\frac{N_1}{(N_1+N_2)} = x_1$ and $\frac{N_2}{(N_1+N_2)} = x_2$, where x_i is the mole fraction of Ideal Gas i. Therefore,

$$\Delta S = k(N_1 + N_2)[\ln(N_1 + N_2) - x_1 \ln N_1 - x_2 \ln N_2] \tag{6.25}$$

$x_1 + x_2 = 1$; therefore,

$$\ln(N_1 + N_2) = (x_1 + x_2)\ln(N_1 + N_2) \tag{6.26}$$

$$\ln N_1 = \ln[x_1(N_1 + N_2)] \tag{6.27}$$

$$\ln N_2 = \ln[x_2(N_1 + N_2)] \tag{6.28}$$

Hence,

$$\Delta S = k(N_1 + N_2)(-x_1 \ln x_1 - x_2 \ln x_2) \tag{6.29}$$

$N_1 + N_2$ becomes the Avogadro constant when the total number of moles is 1. Therefore,

$$k(N_1 + N_2) = R \tag{6.30}$$

Hence,

$$\Delta S = -R(x_1 \ln x_1 + x_2 \ln x_2) \tag{6.31}$$

This result is identical to that derived using $\mathrm{d}S \equiv \frac{\mathrm{d}Q_{\mathrm{in,rev}}}{T}$(Eq. (6.21)) when the total number of moles is 1.

In this section, we have calculated the entropy of mixing and seen that we can obtain the same result using both macroscopic ($\mathrm{d}S \equiv \frac{\mathrm{d}Q_{\mathrm{in,rev}}}{T}$) and microscopic ($S =$

$k \ln \mathcal{W}$) perspectives. A more general discussion of this subject can be found in books on statistical mechanics or statistical thermodynamics.

6.4 Summary

Entropy, S:

Under macroscopic conditions, entropy, S, has the following relationship with the number of possible arrangements (microscopic states) of the system, $\mathcal{W}$:

$$S = k \ln \mathcal{W} \tag{6.1}$$

where k is the Boltzmann constant.

This equation is equivalent to

$$\mathrm{d}S \equiv \frac{\mathrm{d}Q_{\mathrm{in,rev}}}{T}$$

Temperature dependence of entropy:

Changes in molar entropy against temperature can be calculated using the following equation:

$$\Delta \overline{S} = \frac{1}{n}\left[\int_{T_1}^{T_t} \frac{C_P^{(\mathrm{S_I})}}{T}\mathrm{d}T + \frac{L_t}{T_t} + \int_{T_t}^{T_m} \frac{C_P^{(\mathrm{S_{II}})}}{T}\mathrm{d}T + \frac{L_m}{T_m} + \int_{T_m}^{T_e} \frac{C_P^{(\ell)}}{T}\mathrm{d}T + \frac{L_e}{T_e} + \int_{T_e}^{T_2} \frac{C_P^{(\mathrm{g})}}{T}\mathrm{d}T\right] \tag{6.32}$$

where $\mathrm{S_I}$ and $\mathrm{S_{II}}$ are Solid Phases I and II, respectively; L_t, L_m, and L_e are the latent heat at T_t, the heat of melting at T_m, and the heat of vapourisation at T_e, respectively; $C_P^{(\ell)}$ and $C_P^{(\mathrm{g})}$ are the C_P of the liquid phase and the gas phase, respectively; and n is the number of moles.

Chapter 7
Equilibrium Conditions

In this chapter, the conditions for equilibrium states in a closed system with various restrictions will be described based on the general conditions for equilibrium for closed systems (Sect. 7.1.1). In addition, the extremely important state functions Helmholtz energy (Sect. 7.1.4) and Gibbs energy (Sect. 7.1.5) will be defined, and several basic equations for the equilibrium state of a closed system will be derived. The conditions for equilibrium and basic equations shown in Sects. 7.1–7.3 are important with regard to understanding the nature of chemical thermodynamics. Therefore, the reader should fully grasp the contents of these sections. In Sect. 7.5 a condition for heterogeneous equilibrium, the phase rule, is discussed, and in Sect. 7.4 a criterion for equilibrium for a system that involves chemical reactions is discussed.

7.1 Closed Systems

7.1.1 General Conditions for Equilibrium for Closed Systems

Here, we consider a virtual change, from one state to another that is close to the previous state (virtual variation is denoted by δ in some books, i.e. δS and δQ. However, for the sake of simplicity, it is here denoted by d, as in previous chapters. See Sect. 2.1 for the meaning of δ).

In this case, if

$$\mathrm{d}S > \frac{\mathrm{d}Q}{T} \tag{7.1}$$

this change is an irreversible process, according to the second law of thermodynamics (Eq. (5.44)), and will occur spontaneously. However, if

$$\mathrm{d}S \le \frac{\mathrm{d}Q}{T} \tag{7.2}$$

T. Matsushita and K. Mukai, *Chemical Thermodynamics in Materials Science*,
https://doi.org/10.1007/978-981-13-0405-7_7

for all conceivable virtual processes, there is no reversible process in the system and the system must be in an equilibrium state. As there is no macroscopic change in the system, the conditions for thermodynamic equilibrium outlined in Sect. 2.3 are satisfied.

The above equation is a general condition for equilibrium, but is not convenient for practical use. A more convenient form, however, can be derived using the first law of thermodynamics.

The first law of thermodynamics can be expressed as follows for a closed system:

$$\mathrm{d}Q_{\mathrm{in}} = \mathrm{d}U + \mathrm{d}W_{\mathrm{out}} \tag{7.3}$$

By assuming that the only external force acting on the system is a uniform normal pressure (most reaction systems satisfy this assumption), Eq. (7.3) may be rewritten as

$$\mathrm{d}Q_{\mathrm{in}} = \mathrm{d}U + P\mathrm{d}V \tag{7.4}$$

where P is the pressure exerted on the system by its surroundings.

If other forces are acting on a system and external force fields are not negligible, Eq. (7.4) may be rewritten. An extra term must be added for $\mathrm{d}W_{\mathrm{out}}$: the terms $\sigma\mathrm{d}A$ for capillary force and $E\mathrm{d}q_{\mathrm{c}}$ for an electrostatic field, for example, must be added, giving

$$\mathrm{d}U = \mathrm{d}Q_{\mathrm{in}} - P\mathrm{d}V + \sigma\mathrm{d}A + E\mathrm{d}q_{\mathrm{c}} \tag{7.5}$$

where σ is the surface tension, $\mathrm{d}A$ is the change in surface area, E is the potential difference, and $\mathrm{d}q_{\mathrm{c}}$ is the electric charge transferred against the potential difference. $E\mathrm{d}q_{\mathrm{c}}$ is often taken as an example of the effective work of a system in relation to cell reactions.

By substituting Eq. (7.4) into Eq. (7.2) we obtain

$$\mathrm{d}U + P\mathrm{d}V - T\mathrm{d}S \geq 0 \tag{7.6}$$

This is a criterion for equilibrium of a closed system, and can be applied to both homogeneous and heterogeneous systems.

Under more restricted conditions, more useful relationships can be derived from Eq. (7.6). These are discussed in the sections that follow, which define Helmholtz energy, Gibbs energy, and their applications.

7.1.2 Criterion for Equilibrium for a Closed System at Constant U and V ($dU = 0, dV = 0$)

Based on Eq. (7.6), the criterion for equilibrium for a closed system at constant U and V ($\mathrm{d}U = 0, \mathrm{d}V = 0$) is

$$(\mathrm{d}S)_{U,V} \leq 0 \quad (7.7)$$

In order to come to a deeper understanding of this equation, we will discuss it more specifically, considering all of the available paths under constant U and V conditions. If the entropy change of the system, $\mathrm{d}S$, is always zero or negative for such paths, the system cannot be changed spontaneously, and thus is in an equilibrium state.[1]

The criteria for equilibrium for other conditions (Sects. 7.1.3–7.1.5) can be also discussed in a similar manner for constant U and V conditions.

7.1.3 Criterion for Equilibrium for a Closed System at Constant S and V ($dS = 0, dV = 0$)

$$(\mathrm{d}U)_{S,V} \geq 0 \quad (7.9)$$

7.1.4 Criterion for Equilibrium for a Closed System at Constant T and V ($dT = 0, dV = 0$)

$$(\mathrm{d}U)_{T,V} - T(\mathrm{d}S)_{T,V} \geq 0 \quad (7.10)$$

Introducing a new extensive property called Helmholtz energy, F, allows us to describe this equation in a simpler way. Helmholtz energy is defined as follows:

$$F = U - TS \quad (7.11)$$

---

[1]Here, $P\mathrm{d}V = \mathrm{d}W = 0$ as the closed system is kept at constant U and V ($\mathrm{d}U = 0, \mathrm{d}V = 0$), and Eq. (3.3) states that $\mathrm{d}Q = 0$. Hence, this system is de facto an isolated system. A change that does not satisfy the criterion for equilibrium, i.e.

$(\mathrm{d}S)_{U,V} > 0$ (7.8)

is an irreversible process, as can be seen based on Eq. (7.1). Equation (4.3) implies an increase in entropy. Hence, this expression is identical to one of the descriptions of the second law of thermodynamics, which states that "entropy in an isolated system is increased by an irreversible change".

By differentiating Eq. (7.11), we obtain

$$\mathrm{d}F = \mathrm{d}U - T\mathrm{d}S - S\mathrm{d}T \tag{7.12}$$

Hence, at constant T and V ($\mathrm{d}T = 0, \mathrm{d}V = 0$),

$$(\mathrm{d}F)_{T,V} = (\mathrm{d}U)_{T,V} - T(\mathrm{d}S)_{T,V} \tag{7.13}$$

The criterion for equilibrium at constant T and V can thus be reduced to the simplest form using F:

$$(\mathrm{d}F)_{T,V} \geq 0 \tag{7.14}$$

On the other hand, $\mathrm{d}S > \mathrm{d}Q/T$ is a criterion for an irreversible process, i.e. a criterion for a spontaneous change of a system. In a similar manner to $(\mathrm{d}S)_{U,V} \leq 0$, a criterion for a spontaneous change of a system at constant T and V can be described by F as follows:

$$(\mathrm{d}F)_{T,V} < 0 \tag{7.15}$$

Thus, the direction of a change of a system and its equilibrium position at constant T and V can be determined using Helmholtz energy.

7.1.5 Criterion for Equilibrium for a Closed System at Constant *T* and *P* ($dT = 0, dP = 0$)

The criterion for equilibrium for a closed system at constant T and P ($\mathrm{d}T = 0, \mathrm{d}P = 0$) can be expressed by

$$(\mathrm{d}U)_{T,P} + P(\mathrm{d}V)_{T,P} - T(\mathrm{d}S)_{T,P} \geq 0 \tag{7.16}$$

This criterion is important as most experiments and reactions are performed at constant T and P, and so is frequently used in practice.

Equation (7.16) can be reduced to its simplest form by introducing another new extensive property; Gibbs energy, G. Gibbs energy is defined as follows:

$$G = U + PV - TS = H - TS \tag{7.17}$$

By differentiating this equation, we obtain

$$\mathrm{d}G = \mathrm{d}U + P\mathrm{d}V + V\mathrm{d}P - T\mathrm{d}S - S\mathrm{d}T \tag{7.18}$$

Hence, at constant T and P,

$$(\mathrm{d}G)_{T,P} = (\mathrm{d}U)_{T,P} + P(\mathrm{d}V)_{T,P} - T(\mathrm{d}S)_{T,P} \tag{7.19}$$

The criterion for equilibrium at constant T and P can thus be described in a simpler form as

$$(\mathrm{d}G)_{T,P} \geq 0 \tag{7.20}$$

As is the case for $\mathrm{d}F$ above, a criterion for an irreversible process—i.e. a criterion for a spontaneous change of a system—under constant T and P can be described as follows:

$$(\mathrm{d}G)_{T,P} < 0 \tag{7.21}$$

Here, the direction of the change of a system and its equilibrium position at constant T and P can be determined using Gibbs energy.

Let us discuss the meaning of $(\mathrm{d}G)_{T,P} \geq 0$ in greater detail: Assuming that the state changes from State I to another state that is close to State I through a conceivable path at constant T and P. If the *Gibbs energy change* of the system, $(\mathrm{d}G)_{T,P}$, is always zero or positive for any path of the change, the system should be in an equilibrium state.

In addition, G is described as $G = H - TS$. Therefore, $(\mathrm{d}G)_{T,P} = (\mathrm{d}H)_{T,P} - T(\mathrm{d}S)_{T,P}$. Hence, the $(\mathrm{d}G)_{T,P}$ value can be either positive or negative depending on the values of $(\mathrm{d}H)_{T,P}$—an energy term—and $T(\mathrm{d}S)_{T,P}$—an entropy term. For example, even if a reaction is endothermic $\left((\mathrm{d}H)_{T,P} > 0\right)$, $(\mathrm{d}G)_{T,P}$ may become negative if the entropy term, $T(\mathrm{d}S)_{T,P}$, is large enough and $(\mathrm{d}H)_{T,P} < T(\mathrm{d}S)_{T,P}$. As a result, such a reaction occurs spontaneously. In the case of an adiabatic process $(\mathrm{d}H)_{T,P}$ is zero, and again the value of $(\mathrm{d}G)_{T,P}$ can be either positive or negative but depends solely on the value of $(\mathrm{d}S)_{T,P}$. Hence, for an adiabatic process, the change occurs spontaneously when entropy increases $\left((\mathrm{d}S)_{T,P} > 0\right)$.

As we have seen, systems in chemical thermodynamics are discussed alongside heat, work, and changes in mass, and all of this is relatively complex. However, the direction of a reaction and equilibrium position can be determined using Helmholtz energy, F, in the case of constant T and V, and using Gibbs energy, G, for constant T and P.

$-(\mathrm{d}G)_{T,P}$ corresponds to the work done leaving aside work in relation to expansion, i.e. the maximum effective work, $W_{\mathrm{e,max}}$. This relates to the fact that, according to the first law of thermodynamics and as shown in Eq. (7.5), $\mathrm{d}Q_{\mathrm{in}}$ is described as follows, together with effective work, W_{e}:

$$\mathrm{d}Q_{\mathrm{in}} = \mathrm{d}U + P\mathrm{d}V + \mathrm{d}W_{\mathrm{e}} \tag{7.22}$$

when there are other forces acting on a system than P and the external force fields are not negligible. By substituting the equation for the second law of thermodynamics $\left(\mathrm{d}S \geq \frac{\mathrm{d}Q}{T}\right)$, we obtain

$$(\mathrm{d}W_{\mathrm{e}})_{T,P} \leq T(\mathrm{d}S)_{T,P} - (\mathrm{d}U)_{T,P} - P(\mathrm{d}V)_{T,P} = -(\mathrm{d}G)_{T,P} = \mathrm{d}W_{\mathrm{e,max}} \quad (7.23)$$

As regards the electrical work that is done by a system to its surroundings; electrical work is $EZ\mathcal{F}$ when 1 mol of a i^{z+} ion is transferred via a cell reaction for a galvanic cell with electromotive force (E). If the change can be regarded as a reversible process, the maximum effective work, $W_{\mathrm{e,max}}$, is $W_{\mathrm{e,max}} = EZ\mathcal{F} = -\Delta G$ where E is the electromotive force, Z is the charge number of the ion, and $\mathcal{F}$ is the Faraday constant ($= 9.64853 \times 10^4\,\mathrm{C \cdot mol^{-1}}$).

7.1.6 Basic Equations that Are True for a Closed System in Equilibrium

In this section, several basic equations that are true for an equilibrium state of a closed system are derived. According to the second law of thermodynamics, for a reversible process (i.e. a change under equilibrium conditions),

$$\mathrm{d}S = \frac{\mathrm{d}Q_{\mathrm{in,rev}}}{T} \quad (7.24)$$

When there is no force acting on a system except for uniform normal pressure, $\mathrm{d}W_{\mathrm{out}} = P\mathrm{d}V$.

Hence, the differential forms of the first law of thermodynamics are:

$$\mathrm{d}U = T\mathrm{d}S - P\mathrm{d}V \quad (7.25)$$

Based on Eqs. (7.25) and (4.1),

$$\mathrm{d}H = T\mathrm{d}S + V\mathrm{d}P \quad (7.26)$$

Likewise,

$$\mathrm{d}F = -S\mathrm{d}T - P\mathrm{d}V \quad (7.27)$$

$$\mathrm{d}G = -S\mathrm{d}T + V\mathrm{d}P \quad (7.28)$$

By using these four basic equations for $\mathrm{d}U$, $\mathrm{d}H$, $\mathrm{d}F$, and $\mathrm{d}G$ (Eqs. (7.25)–(7.28)), several important relational expressions can be derived. The following differentiated forms are commonly used:

$$\left(\frac{\partial U}{\partial S}\right)_V = T \quad (7.29)$$

$$\left(\frac{\partial U}{\partial V}\right)_S = -P \quad (7.30)$$

$$\left(\frac{\partial G}{\partial T}\right)_P = -S \quad (7.31)$$

$$\left(\frac{\partial G}{\partial P}\right)_T = V \quad (7.32)$$

$$\left(\frac{\partial H}{\partial S}\right)_P = T \quad (7.33)$$

$$\left(\frac{\partial H}{\partial P}\right)_S = V \quad (7.34)$$

$$\left(\frac{\partial F}{\partial T}\right)_V = -S \quad (7.35)$$

$$\left(\frac{\partial F}{\partial V}\right)_T = -P \quad (7.36)$$

Based on Eq. (7.31), we know that the temperature dependence of G, i.e. the temperature coefficient of G, corresponds to $-S$. Entropy, S, is always positive except for at 0 K. Hence, G decreases with increasing temperature. In addition, the entropy of a liquid is higher than that of a solid, and the entropy of a gas is higher than that of a liquid. Hence, the degree by which G decreases with temperature is as follows: solid < liquid < gas. Together with enthalpy change by temperature $\left(\left(\frac{\partial H}{\partial T}\right)_P = C_P\right)$, change in G is also shown for a pure material in Fig. 7.1.

In addition to Eqs. (7.29)–(7.32), Maxwell's relations are also often used.

For the function

$$z = \mathrm{f}(x, y) \quad (7.37)$$

the total derivative is described as follows:

$$\mathrm{d}z = \left(\frac{\partial \mathrm{f}}{\partial x}\right)_y + \left(\frac{\partial \mathrm{f}}{\partial y}\right)_x \quad (7.38)$$

If the total derivative is an exact differential, the following relationship is satisfied:

$$\left[\frac{\partial}{\partial y}\left(\frac{\partial \mathrm{f}}{\partial x}\right)_y\right]_x = \left[\frac{\partial}{\partial x}\left(\frac{\partial \mathrm{f}}{\partial y}\right)_x\right]_y \quad (7.39)$$

Mathematically, the total derivative of the state functions (U, H, F, and G) is the exact differential.

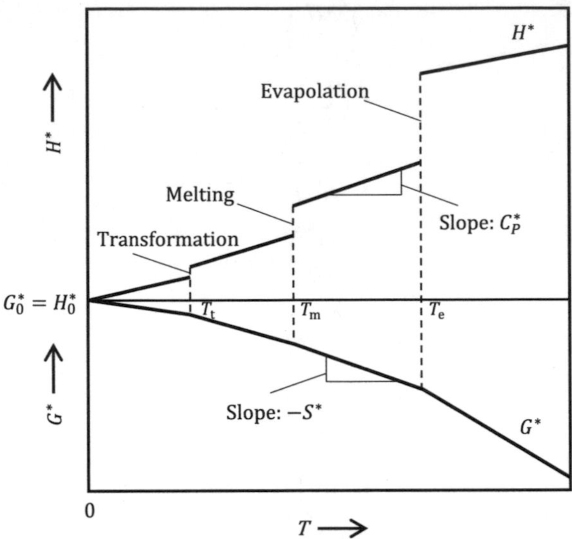

Fig. 7.1 The relationship between G^*, H^*, and temperature

Therefore, using Eqs. (7.29) and (7.30),

$$\left(\frac{\partial T}{\partial V}\right)_S = -\left(\frac{\partial P}{\partial S}\right)_V \tag{7.40}$$

In a similar manner, the following relationships can be derived:

$$\left(\frac{\partial V}{\partial T}\right)_P = -\left(\frac{\partial S}{\partial P}\right)_T \tag{7.41}$$

$$\left(\frac{\partial T}{\partial P}\right)_S = \left(\frac{\partial V}{\partial S}\right)_P \tag{7.42}$$

$$\left(\frac{\partial P}{\partial T}\right)_V = \left(\frac{\partial S}{\partial V}\right)_T \tag{7.43}$$

Equations (7.40)–(7.43) are known as *Maxwell's relations*. The quantities on the left-hand sides of these equations are experimentally measurable, and so it is possible to ascertain the value of the non-measurable quantities (those on the right-hand sides of the equations) using Maxwell's relations.

7.2 Open Systems

7.2.1 Basic Equations for Homogeneous Open Systems and Chemical Potential

In this section, open systems, which are one of the most important types of system, are considered, and several basic equations relating to homogeneous open systems are derived.

When we apply the first law of thermodynamics to an open system, the mass term must be added to the relevant equation. Hence, the criterion for equilibrium for a homogeneous open system that does not involve chemical reactions is

$$(\mathrm{d}U)_{S,V,n_i} \geq 0 \tag{7.44}$$

By keeping mass (n_i) constant, the criterion for equilibrium for a closed system can be applied.

As internal energy, U, is an extensive property (as is discussed in Sect. 2.3), it is as follows in an equilibrium state when $r + 2$ variables, including V and S, are chosen.

$$U = U(V, S, n_i, \cdots n_r) \tag{7.45}$$

where n_i is the number of moles of component i.

Hence, the total derivative of U is

$$\mathrm{d}U = \left(\frac{\partial U}{\partial V}\right)_{S,n_i} \mathrm{d}V + \left(\frac{\partial U}{\partial S}\right)_{V,n_i} \mathrm{d}S + \sum_{i=1}^{r} \left(\frac{\partial U}{\partial n_i}\right)_{S,V,n_j(j=1,\cdots,r,j\neq i)} \mathrm{d}n_i \quad (i = 1, \ldots, r) \tag{7.46}$$

The following equations are true, as the basic equations for a closed system (Eq. (7.25)) can be applied at constant n_i:

$$\left(\frac{\partial U}{\partial S}\right)_{V,n_i} = T \tag{7.47}$$

$$\left(\frac{\partial U}{\partial V}\right)_{S,n_i} = -P \tag{7.48}$$

Hence,

$$\mathrm{d}U = T\mathrm{d}S - P\mathrm{d}V + \sum_{i=1}^{r} \mu_i \mathrm{d}n_i \tag{7.49}$$

μ_i is defined as

$$\mu_i = \left(\frac{\partial U}{\partial n_i}\right)_{S,V,n_j(j=1,\cdots,r,j\neq i)} \tag{7.50}$$

and μ_i is the *chemical potential* of component i.

Based on $\mathrm{d}H = \mathrm{d}U + P\mathrm{d}V + V\mathrm{d}P$ and Eq. (7.49),

$$\mathrm{d}H = T\mathrm{d}S + V\mathrm{d}P + \sum_{i=1}^{r} \mu_i \mathrm{d}n_i \tag{7.51}$$

On the other hand, the total derivative of enthalpy, $H(S, P, n_i)$, is

$$\mathrm{d}H = T\mathrm{d}S + V\mathrm{d}P + \sum_{i=1}^{r} \left(\frac{\partial H}{\partial n_i}\right)_{S,P,n_j(j=1,\ldots,r,j\neq i)} \mathrm{d}n_i \tag{7.52}$$

Therefore, by comparing Eqs. (7.51) and (7.52), we obtain

$$\mu_i = \left(\frac{\partial U}{\partial n_i}\right)_{S,V,n_j(j=1,\ldots,r,j\neq i)} = \left(\frac{\partial H}{\partial n_i}\right)_{S,P,n_j(j=1,\ldots,r,j\neq i)} \tag{7.53}$$

Likewise, based on $\mathrm{d}F = \mathrm{d}U - T\mathrm{d}S - S\mathrm{d}T$ and Eq. (7.49),

$$\mathrm{d}F = -S\mathrm{d}T + P\mathrm{d}V + \sum_{i=1}^{r} \mu_i \mathrm{d}n_i \tag{7.54}$$

The total derivative of $F(T, V, n_i)$ is

$$\mathrm{d}F = -S\mathrm{d}T - P\mathrm{d}V + \sum_{i=1}^{r} \left(\frac{\partial F}{\partial n_i}\right)_{T,V,n_j(j=1,\ldots,r,j\neq i)} \mathrm{d}n_i \tag{7.55}$$

Therefore, by comparing Eqs. (7.54) and (7.55), we obtain

$$\mu_i = \left(\frac{\partial U}{\partial n_i}\right)_{S,V,n_j(j=1,\ldots,r,j\neq i)} = \left(\frac{\partial F}{\partial n_i}\right)_{T,V,n_j(j=1,\ldots,r,j\neq i)} \tag{7.56}$$

Likewise, based on $\mathrm{d}G = \mathrm{d}H - T\mathrm{d}S - S\mathrm{d}T$ and Eq. (7.51),

$$\mathrm{d}G = -S\mathrm{d}T + V\mathrm{d}P + \sum_{i=1}^{r} \mu_i \mathrm{d}n_i \tag{7.57}$$

The total derivative of $G(T, P, n_i)$ is

$$\mathrm{d}G = -S\mathrm{d}T - V\mathrm{d}P + \sum_{i=1}^{r} \left(\frac{\partial G}{\partial n_i}\right)_{T,P,n_j(j=1,\ldots,r,j\neq i)} \mathrm{d}n_i \tag{7.58}$$

Therefore, by comparing Eqs. (7.57) and (7.58), we obtain

$$\mu_i = \left(\frac{\partial U}{\partial n_i}\right)_{S,V,n_j(j=1,\ldots,r,j\neq i)} = \left(\frac{\partial G}{\partial n_i}\right)_{T,P,n_j(j=1,\ldots,r,j\neq i)} \tag{7.59}$$

Based on Eq. (7.59), the chemical potential of component i in a solution, μ_i, can be regarded as a Gibbs energy change of the solution when 1 mol of component i is added at constant T and P (assuming the change in concentration of the solution is negligible).

In summary, we have obtained the following relationships:

$$\mu_i = \left(\frac{\partial U}{\partial n_i}\right)_{S,V,n_j(j=1,\ldots,r,j\neq i)} \tag{7.60}$$

$$= \left(\frac{\partial H}{\partial n_i}\right)_{S,P,n_j(j=1,\ldots,r,j\neq i)} \tag{7.61}$$

$$= \left(\frac{\partial F}{\partial n_i}\right)_{T,V,n_j(j=1,\ldots,r,j\neq i)} \tag{7.62}$$

$$= \left(\frac{\partial G}{\partial n_i}\right)_{T,P,n_j(j=1,\ldots,r,j\neq i)} \tag{7.63}$$

The relational expressions between $\mathrm{d}U$, $\mathrm{d}H$, $\mathrm{d}F$, $\mathrm{d}G$, and μ_i Eqs. (7.60)–(7.63) are the basic equations for a homogeneous open system under equilibrium conditions. Chemical potential, μ_i, which was defined by J. W. Gibbs, is a partial molar quantity (e.g. partial molar Gibbs energy) and an extensive property.

7.2.2 The Gibbs-Duhem Equation

As can be seen in Eq. (7.59), Gibbs energy can be expressed by the following equation:

$$G = \sum_{i=1}^{r} n_i \mu_i \tag{7.64}$$

where n_i is the number of moles of component i and μ_i is the chemical potential (partial molar Gibbs energy) of component i.

Differentiating this equation results in

$$\mathrm{d}G = \sum_{i=1}^{r} \mu_i \mathrm{d}n_i + \sum_{i=1}^{r} n_i \mathrm{d}\mu_i \tag{7.65}$$

Based on Eq. (7.65)—which states that G is a function of μ_i and n_i—and the basic equation for G for homogeneous open systems (Eq. (7.57)), G is a function of T, P and n_i:

$$\sum_{i=1}^{r} n_i \mathrm{d}\mu_i + S\mathrm{d}T - V\mathrm{d}P = 0 \tag{7.66}$$

By dividing both sides of the equation by the total mole number, $n = \sum_{i=1}^{r} n_i$, we obtain

$$\sum_{i=1}^{r} x_i \mathrm{d}\mu_i + \overline{S}\mathrm{d}T - \overline{V}\mathrm{d}P = 0 \tag{7.67}$$

where $\overline{S}$ and $\overline{V}$ are the mean molar entropy and the mean molar volume, respectively.

This equation was derived by Gibbs and Duhem independently, and is called the *Gibbs-Duhem equation.*

At constant T and P, the equation becomes

$$\sum_{i=1}^{r} n_i \mathrm{d}\mu_i = \sum_{i=1}^{\mathrm{r}} x_i \mathrm{d}\mu_i = 0 \tag{7.68}$$

The Gibbs-Duhem equation is important as it can be used to calculate activity (see Sect. 8.4.2).

7.3 Isolated and Heterogeneous Systems

7.3.1 Criterion for Equilibrium for an Isolated Heterogeneous System

In this section, the criterion for equilibrium for an isolated *heterogeneous system* that does not involve chemical reactions is presented. In most cases, a system becomes an isolated heterogeneous system if we extend it to a certain extent (cf. Fig. 2.2). Therefore, the relational expressions for extensive properties, which are derived as criteria for the equilibrium of such a system, are necessary in order to derive the phase rule, to handle partition equilibrium between different phases, and so on.

Next, let us discuss isolated heterogeneous systems.

In a heterogeneous system, there are several phases (at least two phases). It is assumed that individual phases themselves are in equilibrium and isolated from one another. If we can find the necessary and sufficient conditions to ensure the equilibrium for each phase in relation to the extensive properties of a heterogeneous system, and if this is the case even if we remove the 'isolated' constraint, the necessary and sufficient conditions are equivalent to those of a heterogeneous system.

The criteria for closed systems were derived in the earlier sections of this book. Thus, we will temporarily assume that the system under discussion is a closed system, and apply the criteria for such a system. The subsequent discussion, however, shows that such a system is a de facto isolated system.

Just as in Eq. (7.44), the criterion for equilibrium using internal energy is

$$(\mathrm{d}U)_{S,V,n_i} \geq 0 \tag{7.69}$$

Regarding the internal energy of a system as the summation of the internal energies of each phase is perfectly acceptable, as internal energy is an extensive property and the influence of the surface of the phase(s) can be ignored due to the fact that the system consists of a large quantity of phases.

The entropy and mass of each component are also extensive properties, and as such can also be treated as the summation of the properties of each phase. Hence, for a system that consists of ν phases (bulk phase), the following relationships are true:

$$\begin{aligned} U &= \sum_{\alpha=1}^{\nu} U^{(\alpha)} \\ S &= \sum_{\alpha=1}^{\nu} S^{(\alpha)} \\ V &= \sum_{\alpha=1}^{\nu} V^{(\alpha)} \\ n_i &= \sum_{\alpha=1}^{\nu} n_i^{(\alpha)}, \quad i = 1, \ldots\ldots, r \end{aligned} \tag{7.70}$$

where (α) denotes the thermodynamic properties of the α phase.

The criterion $(\mathrm{d}U)_{S,V,n_i} \geq 0$ can be applied to the change at constant S, V, and $n_i (i = 1, \ldots\ldots, r)$. Therefore, as a matter of course, the following relationships must be satisfied (i.e. the path of the virtual change in internal energy must satisfy these conditions):

$$\mathrm{d}S = \sum_{\alpha=1}^{\nu} \mathrm{d}S^{(\alpha)} = 0 \tag{7.71}$$

$$\mathrm{d}V = \sum_{\alpha=1}^{\nu} \mathrm{d}V^{(\alpha)} = 0 \tag{7.72}$$

$$\mathrm{d}n_i = \sum_{\alpha=1}^{\nu} \mathrm{d}n_i^{(\alpha)} = 0, \quad i = 1, \ldots\ldots, r \tag{7.73}$$

$\mathrm{d}V = 0$ implies that $\mathrm{d}W_{\mathrm{out}}(= P\mathrm{d}V) = 0$. $\mathrm{d}n_i = 0$ means that there is no exchange of mass between a system and its surroundings. In addition, there is no internally produced entropy as it is assumed that each phase is in an equilibrium state (that is, a reversible process). Furthermore, the total entropy of a heterogeneous system is conservative ($\mathrm{d}S = 0$). Hence, the above-mentioned restriction implies that the system is practically isolated and does not involve chemical reactions.

Now, let us assume that each homogeneous phase in a heterogeneous system is in equilibrium and isolated from every other phase. Let us also assume that each phase undergoes virtual change, $\mathrm{d}U^{(\alpha)}$, while maintaining equilibrium under $\mathrm{d}S = 0$, $\mathrm{d}V = 0$, and $\mathrm{d}n_i = 0$ conditions when the individual phases come into contact with one another. As is discussed above, the criterion for equilibrium for a heterogeneous system is $(\mathrm{d}U)_{S,V,n_i} \geq 0$.

It should be noted that, if each homogeneous phase is removed from a heterogeneous system, the system can be treated as an open homogeneous system. Therefore, the following equation can be applied for $\mathrm{d}U^{(\alpha)}$:

$$\mathrm{d}U^{(\alpha)} = T^{(\alpha)}\mathrm{d}S^{(\alpha)} - P^{(\alpha)}\mathrm{d}V^{(\alpha)} + \sum_{i=1}^{r} \mu_i^{(\alpha)}\mathrm{d}n_i^{(\alpha)} \tag{7.74}$$

Hence, the criterion for equilibrium for the system is

$$\mathrm{d}U = \sum_{\alpha=1}^{\nu}\left(T^{(\alpha)}\mathrm{d}S^{(\alpha)} - P^{(\alpha)}\mathrm{d}V^{(\alpha)} + \sum_{i=1}^{r} \mu_i^{(\alpha)}\mathrm{d}n_i^{(\alpha)}\right) \geq 0 \tag{7.75}$$

The total entropy, volume, and mass of a heterogeneous system are conservative ($\mathrm{d}S = 0$, $\mathrm{d}V = 0$, and $\mathrm{d}n_i = 0$, $(i = 1, \ldots\ldots, r)$).

The above equation constitutes a criterion that maintains the equilibrium of a heterogeneous system after contact has occurred between individual isolated phases.

Now, we will derive the conditions for the intensive properties $T^{(\alpha)}$, $P^{(\alpha)}$, and $\mu_i^{(\alpha)}(i = 1, \ldots\ldots, r$, and $\alpha = 1, \ldots\ldots, \nu)$. In order to accomplish this, we will consider several possible variations ($\mathrm{d}X$) for the extensive properties S, $V^{(\alpha)}$, and $n_i^{(\alpha)}$ in specific circumstances.

As an example, consider the relationship of pressure: The variations of the extensive properties $S^{(\alpha)}$, $V^{(\alpha)}$, and $n_i^{(\alpha)}(i = 1, \ldots\ldots, r$, and $\alpha = 1, \ldots\ldots, \nu)$ are as follows:

$$\mathrm{d}S^{(\alpha)} = 0, \ (\alpha = 1, \ldots\ldots, \nu) \tag{7.76}$$

$$\mathrm{d}n_i^{(\alpha)} = 0, \ (i = 1, \ldots\ldots, r, \text{ and } \alpha = 1, \ldots\ldots, \nu) \tag{7.77}$$

$$\mathrm{d}V^{(\alpha)} = 0, \ (\alpha = 3, \ldots\ldots, \nu) \tag{7.78}$$

$$\mathrm{d}V^{(1)} = -\mathrm{d}V^{(2)} \tag{7.79}$$

In this case, the criterion for equilibrium is

$$dU = \sum_{\alpha=1}^{\nu} \left(T^{(\alpha)} dS^{(\alpha)} - P^{(\alpha)} dV^{(\alpha)} + \sum_{i=1}^{r} \mu_i^{(\alpha)} dn_i^{(\alpha)} \right) \geq 0 \tag{7.80}$$

$dS^{(\alpha)} = 0$ and $dn_i^{(\alpha)} = 0$; thus,

$$dU = \sum_{\alpha=1}^{\nu} \left(-P^{(\alpha)} dV^{(\alpha)} \right) \geq 0 \tag{7.81}$$

$dV^{(\alpha)} = 0, (\alpha = 3, \ldots\ldots, \nu)$,
Hence,

$$dU = -P^{(1)} dV^{(1)} - P^{(2)} dV^{(2)} \geq 0 \tag{7.82}$$

$dV^{(1)} = -dV^{(2)}$; thus,

$$dU = \left(P^{(2)} - P^{(1)} \right) dV^{(1)} \geq 0 \tag{7.83}$$

(a) If the two phases are separated by a deformable interface, the value of $dV^{(1)}$ can be either positive or negative, and must be $P^{(2)} - P^{(1)} = 0$ to satisfy the criterion stated in Eq. (7.83). If all of the interfaces between the phases are deformable, $P^{(2)} = P^{(3)}$, $P^{(3)} = P^{(4)}, \ldots\ldots$ are obtained in a similar manner. As a result, the following condition must be satisfied for this heterogeneous system:

$$P^{(1)} = P^{(2)} = P^{(3)} = \ldots\ldots = P^{(\nu)} \tag{7.84}$$

(b) If $dV^{(1)} \geq 0$, i.e. if Phase 1 is unable to reduce in volume, the pressure of the two phases must satisfy the following criterion in order to establish equilibrium:

$$P^{(2)} - P^{(1)} \geq 0 \tag{7.85}$$

(c) If the interface between Phases 1 and 2 is a rigid body, $dV^{(1)} = 0$. Therefore, there is no restriction on pressure.

Having discussed the conditions for pressure, we can now turn to temperature, T, and chemical potential, μ_i.

In the case of pressure, dV is taken to be a variation. In place of dV in the above example, dS for temperature (T) and dn_i for chemical potential (μ_i) can be taken as variations. The following conditions will then be obtained:
If,

(1) all interfaces are deformable,
(2) all interfaces are heat conductors, and
(3) all interfaces are pervious to all of the components,

the following conditions to establish equilibrium for a heterogeneous system are obtained:

$$P^{(\alpha)} = P \text{ for all } \alpha \tag{7.86}$$

$$T^{(\alpha)} = T \text{ for all } \alpha \tag{7.87}$$

$$\mu_i^{(\alpha)} = \mu_i \text{ for all } \alpha \text{ and } i\,(i = 1, \ldots\ldots, r) \tag{7.88}$$

By substituting these conditions (Eqs. (7.86)–(7.88)) into Eq. (7.80), we obtain $(\mathrm{d}U)_{S,V,n_i} = 0$, which is a sufficient condition to establish the equilibrium.

Note that all assumptions made during the derivation must be satisfied to apply the conditions for equilibrium of the system to P, T, and μ_i . In most cases, these assumptions are satisfied, but there are some well-known exceptions. For example:

(1) Equation (7.70) is not satisfied, i.e. the influence of the surface cannot be ignored. If Phase α is a fine particle, its internal pressure, $P^{(\alpha)}$, is equal to the external pressure, $P^{(\beta)}$, and the relationship is as follows:

$$P^{(\alpha)} - P^{(\beta)} = \frac{2\sigma}{r} \tag{7.89}$$

where σ and r are the surface tension and the radius of the fine particle, respectively.

(2) The interface is semipermeable, i.e. it is impervious to component i.
In this case, there are no restrictions on the chemical potentials, μ_i, of either side of the interface.

Example 7.1

Derive the necessary condition for μ_i required to establish the equilibrium of a heterogeneous system:

$\mu_i^{(1)} = \mu_i^{(2)} = \ldots\ldots = \mu_i$ Eq. (7.88)

Refer to the derivation for the condition of P.

Solution:

The variations of the extensive properties $S^{(\alpha)}, V^{(\alpha)}$, and $n_i^{(\alpha)} (i = 1, \ldots\ldots, r, \text{ and } \alpha = 1, \ldots\ldots, \nu)$ are as follows:

$$\mathrm{d}S^{(\alpha)} = 0, \quad (\alpha = 1, \ldots\ldots, \nu)$$

$$\mathrm{d}V^{(\alpha)} = 0, \quad (\alpha = 1, \ldots\ldots, r)$$

$$\mathrm{d}n_j^{(\alpha)} = 0, \quad (j = 1, \ldots\ldots, r, \text{ and } \alpha = 1, \ldots\ldots, \nu)(j \neq i)$$

$$\mathrm{d}n_i^{(\alpha)} = 0, \quad (\alpha = 3 \ldots\ldots, \nu)$$

$$\mathrm{d}n_i^{(1)} = -\mathrm{d}n_i^{(2)}$$

In this case, the criterion for equilibrium is

$$\mathrm{d}U = \sum_{\alpha=1}^{\nu} \left(T^{(\alpha)}\mathrm{d}S^{(\alpha)} - P^{(\alpha)}\mathrm{d}V^{(\alpha)} + \sum_{i=1}^{r} \mu_i^{(\alpha)}\mathrm{d}n_i^{(\alpha)} \right) \geq 0$$

$\mathrm{d}S^{(\alpha)} = 0$ and $\mathrm{d}V^{(\alpha)} = 0$, thus

$$\mathrm{d}U = \sum_{\alpha=1}^{\nu} \mathrm{d}U^{(\alpha)} = \sum_{\alpha=1}^{\nu}\sum_{i=1}^{r} \mu_i^{(\alpha)}\mathrm{d}n_i^{(\alpha)} \geq 0$$

Hence,

$$\mu_i^{(1)}\mathrm{d}n_i^{(1)} + \mu_i^{(2)}\mathrm{d}n_i^{(2)} \geq 0$$

Thus,

$$\left(\mu_i^{(1)} - \mu_i^{(2)}\right)\mathrm{d}n_i^{(1)} \geq 0$$

Therefore, it must be the case that $\mu_i^{(1)} = \mu_i^{(2)}$ for the above equation to be satisficd. The relationships $\mu_i^{(2)} = \mu_i^{(3)}, \ldots\ldots$ can be obtained in a similar manner. As a result, the following condition must be satisfied for equilibrium in a heterogeneous system:

$$\mu_i^{(1)} = \mu_i^{(2)} = \mu_i^{(3)} = \ldots\ldots = \mu_i^{(\nu)} = \mu_i$$

7.3.2 *The Influence of Temperature and Pressure on Melting Phenomena (The Clausius-Clapeyron Equation)*

A practical issue relating to heterogeneous systems is the phase transformation of unary systems, e.g. phase transformation of solids, melting, and evaporation phenomena. The influence of temperature and pressure on these is discussed below.

According to the condition for equilibrium for a heterogeneous system (Eq. (7.88)),

$$\mu_i^{*,(\mathrm{s})}(T_\mathrm{m}, P) = \mu_i^{*,(\ell)}(T_\mathrm{m}, P) \tag{7.90}$$

where T_{m} is the melting point of pure component i under pressure P, $\mu_i^{*,(\mathrm{s})}$ is the chemical potential of pure solid i, and $\mu_i^{*,(\ell)}$ is the chemical potential of pure liquid i. The superscript * denotes a pure (unary) system.

Based on Eqs. (7.28) ($\mathrm{d}G = -S\mathrm{d}T + V\mathrm{d}P$) and (7.63) $\left(\mu_i = \left(\frac{\partial G}{\partial n_i}\right)_{T,P,n_j(j=1,\cdots,r,j\neq i)}\right)$, a change in chemical potential, $\mathrm{d}\mu_i^*$, can be described as follows, wherein a change in melting point, $\mathrm{d}T_{\mathrm{m}}$ occurs as a result of a change in pressure, $\mathrm{d}P$:

$$\mathrm{d}\mu_i^{*,(\mathrm{s})} = -\overline{S}_i^{*,(\mathrm{s})}\mathrm{d}T_{\mathrm{m}} + \overline{V}_i^{*,(\mathrm{s})}\mathrm{d}P \tag{7.91}$$

$$\mathrm{d}\mu_i^{*,(\ell)} = -\overline{S}_i^{*,(\ell)}\mathrm{d}T_{\mathrm{m}} + \overline{S}_i^{*,(\ell)}\mathrm{d}P \tag{7.92}$$

The phases are in equilibrium even after the change, $\mathrm{d}T_{\mathrm{m}}$ and $\mathrm{d}P$. Therefore,

$$\mu_i^{*,(\mathrm{s})}(T_{\mathrm{m}} + \mathrm{d}T_{\mathrm{m}},\, P + \mathrm{d}P) = \mu_i^{*,(\ell)}(T_{\mathrm{m}} + \mathrm{d}T_{\mathrm{m}},\, P + \mathrm{d}P) \tag{7.93}$$

Each side of the above equation may be reduced as follows:

$$\mu_i^{*,(\mathrm{s})}(T_{\mathrm{m}} + \mathrm{d}T_{\mathrm{m}},\, P + \mathrm{d}P) = \mu_i^{*,(\mathrm{s})} + \mathrm{d}\mu_i^{*,(\mathrm{s})} = \mu_i^{*,(\mathrm{s})}(T_{\mathrm{m}},\, P) - \overline{S}_i^{*,(\mathrm{s})}\mathrm{d}T_{\mathrm{m}} + \overline{V}_i^{*,(\mathrm{s})}\mathrm{d}P \tag{7.94}$$

$$\mu_i^{*,(\ell)}(T_{\mathrm{m}} + \mathrm{d}T_{\mathrm{m}},\, P + \mathrm{d}P) = \mu_i^{*,(\ell)} + \mathrm{d}\mu_i^{*,(\ell)} = \mu_i^{*,(\ell)}(T_{\mathrm{m}},\, P) - \overline{S}_i^{*,(\ell)}\mathrm{d}T_{\mathrm{m}} + \overline{V}_i^{*,(\ell)}\mathrm{d}P \tag{7.95}$$

Based on Eqs. (7.90), (7.93), (7.94), and (7.95),

$$-\overline{S}_i^{*,(\mathrm{s})}\mathrm{d}T_{\mathrm{m}} + \overline{V}_i^{*,(\mathrm{s})}\mathrm{d}P = -\overline{S}_i^{*,(\ell)}\mathrm{d}T_{\mathrm{m}} + \overline{V}_i^{*,(\ell)}\mathrm{d}P \tag{7.96}$$

Therefore,

$$\frac{\mathrm{d}T_{\mathrm{m}}}{\mathrm{d}P} = \frac{\overline{V}_i^{*,(\ell)} - \overline{V}_i^{*,(\mathrm{s})}}{\overline{S}_i^{*,(\ell)} - \overline{S}_i^{*,(\mathrm{s})}} \tag{7.97}$$

By describing a change in thermodynamic properties due to melting using the symbol Δ_{m}, a change in chemical potential can be described as follows:

$$\Delta_{\mathrm{m}}\mu_i^* = \mu_i^{*,(\ell)} - \mu_i^{*,(\mathrm{s})} \tag{7.98}$$

$G = H - TS$, therefore $\mu_i^{*,(\ell)} = \overline{H}_i^{*,(\ell)} - T_{\mathrm{m}}\overline{S}_i^{*,(\ell)}$ and $\mu_i^{*,(\mathrm{s})} = \overline{H}_i^{*,(\mathrm{s})} - T_{\mathrm{m}}\overline{S}_i^{*,(\mathrm{s})}$. Hence,

$$
\begin{aligned}
\Delta_{\mathrm{m}}\mu_i^* &= \mu_i^{*,(\ell)} - \mu_i^{*,(\mathrm{s})} \\
&= \overline{H}_i^{*,(\ell)} - T_{\mathrm{m}}\overline{S}_i^{*,(\ell)} - \left(\overline{H}_i^{*,(\mathrm{s})} - T_{\mathrm{m}}\overline{S}_i^{*,(\mathrm{s})}\right) \\
&= \Delta_{\mathrm{m}}\overline{H}_i^* - T_{\mathrm{m}}\left(\overline{S}_i^{*,(\ell)} - \overline{S}_i^{*,(\mathrm{s})}\right)
\end{aligned} \tag{7.99}
$$

As shown in Eq. (7.93),

$$
\Delta_{\mathrm{m}}\mu_i^* = \mu_i^{*,(\ell)}(T_{\mathrm{m}}, P) - \mu_i^{*,(\mathrm{s})}(T_{\mathrm{m}}, P) = 0 \tag{7.100}
$$

Hence,

$$
\Delta_{\mathrm{m}}\mu_i^* = \Delta_{\mathrm{m}}\overline{H}_i^* - T_{\mathrm{m}}\left(\overline{S}_i^{*,(\ell)} - \overline{S}_i^{*,(\mathrm{s})}\right) = 0 \tag{7.101}
$$

$$
\left(\overline{S}_i^{*,(\ell)} - \overline{S}_i^{*,(\mathrm{s})}\right) = \frac{\Delta_{\mathrm{m}}\overline{H}_i^*}{T_{\mathrm{m}}} \tag{7.102}
$$

By substituting this equation into Eq. (7.97), the following equation is obtained:

$$
\frac{\mathrm{d}T_{\mathrm{m}}}{\mathrm{d}P} = \frac{T_{\mathrm{m}}\Delta_{\mathrm{m}}\overline{V}_i^*}{\Delta_{\mathrm{m}}H_i^*} \tag{7.103}
$$

where $\Delta_{\mathrm{m}}\overline{V}_i^* = \overline{V}_i^{*,(\ell)} - \overline{V}_i^{*,(\mathrm{s})}$ is the molar volume change due to melting and $\Delta_{\mathrm{m}}\overline{H}_i^*$ is the molar enthalpy change of melting, corresponding to the heat of melting per mole. A similar relationship with Eq. (7.103) can be derived for the other phase transformations (phase transformation of a solid, evaporation, etc.). Equation (7.103) is known as the *Clausius-Clapeyron equation*, and is as follows:

$$
\frac{\mathrm{d}T}{\mathrm{d}P} = \frac{T\Delta\overline{V}^*}{\Delta\overline{H}^*} \tag{7.104}
$$

where T and P are the temperature and the pressure when Phases 1 and 2 are in equilibrium, $\Delta\overline{V}^* = \overline{V}^{*,(2)} - \overline{V}^{*,(1)}$ is the molar volume change accompanying the phase transformation, and $\Delta\overline{H}^* = \overline{H}^{*,(2)} - \overline{H}^{*,(1)}$ is the phase transition heat per mole (molar latent heat).

Example 7.2

The thermodynamic properties in relation to the melting of copper at 10^5Pa are as follows:

$$
T_{\mathrm{m}} = 1356\,\mathrm{K}
$$

$$
\Delta_{\mathrm{m}}\overline{H}_{\mathrm{Cu}}^* = 13.05 \times 10^3\mathrm{J}\cdot\mathrm{mol}^{-1}
$$

$$\Delta_{\mathrm{m}}\overline{V}^{*}_{\mathrm{Cu}} = 3.19 \times 10^{-7}\mathrm{m}^{3} \cdot \mathrm{mol}^{-1}$$

Calculate the melting point of copper under 10^{8}Pa.
Assume that the influence of pressure on $\Delta_{\mathrm{m}}\overline{H}^{*}_{\mathrm{Cu}}$ and $\Delta_{\mathrm{m}}\overline{V}^{*}_{\mathrm{Cu}}$ is negligible.
Solution:

$$\frac{\mathrm{d}T}{\mathrm{d}P} = \frac{T_{\mathrm{m}}\Delta_{\mathrm{m}}\overline{V}^{*}}{\Delta_{\mathrm{m}}\overline{H}^{*}}$$

Thus,

$$\int \frac{\mathrm{d}T}{T_{\mathrm{m}}} = \int \frac{\Delta_{\mathrm{m}}\overline{V}^{*}}{\Delta_{\mathrm{m}}\overline{H}^{*}}\mathrm{d}P$$

And so,

$$\ln\frac{T_{\mathrm{m},2}}{T_{\mathrm{m},1}} = \frac{\Delta_{\mathrm{m}}\overline{V}^{*}}{\Delta_{\mathrm{m}}\overline{H}^{*}}(P_2 - P_1)$$

Hence,

$$\ln\frac{T_{\mathrm{m},2}}{1356} = \frac{3.19 \times 10^{-7}}{13.05 \times 10^{3}}\left(10^{8} - 10^{5}\right) = 2.442 \times 10^{-3}\mathrm{m}^{3} \cdot \mathrm{Pa} \cdot \mathrm{J}^{-1}$$

As regards the units: $\mathrm{m}^{3} \cdot \mathrm{Pa} \cdot \mathrm{J}^{-1} = \mathrm{m}^{3} \cdot \left(\mathrm{N} \cdot \mathrm{m}^{-1}\right) \cdot \left(\mathrm{N}^{-1} \cdot \mathrm{m}^{-2}\right) = 1$ (dimensionless).
Hence, the melting point of copper at 10^{8} Pa, $T_{\mathrm{m},2} = 1359\,\mathrm{K}$.

7.3.3 The Influence of Temperature and Pressure on the Conditions for Diamond Synthesis

Both graphite and diamond consist of the same element; carbon. They simply represent two different phases of its solid state. Beginning in the 1920s, multiple attempts were made to synthesise diamonds using graphite or other forms of carbon, and General Electric (GE) successfully produced a synthetic diamond in 1955.

Here, the conditions for the production of a synthetic diamond using thermodynamics are discussed. The reaction by which a synthetic diamond is created can be expressed as follows:

$$\mathrm{C}^{(\mathrm{G})} = \mathrm{C}^{(\mathrm{D})} \tag{7.105}$$

where G and D denote the graphite phase and the diamond phase, respectively.

When dn moles of diamond are synthesised from graphite using the above equation at a constant temperature and pressure, dn can be described as follows for a system that consists of $n_{\mathrm{C}}^{(\mathrm{G})}$ moles of graphite and $n_{\mathrm{C}}^{(\mathrm{D})}$ moles of diamond:

$$\mathrm{d}n = -\mathrm{d}n_{\mathrm{C}}^{(\mathrm{G})} = \mathrm{d}n_{\mathrm{C}}^{(\mathrm{D})} \tag{7.106}$$

According to Eq. (7.64), the total Gibbs energy before the reaction may be expressed as follows:

$$G = n_{\mathrm{C}}^{(\mathrm{G})}\mu_{\mathrm{C}}^{*,(\mathrm{G})} + n_{\mathrm{C}}^{(\mathrm{D})}\mu_{\mathrm{C}}^{*,(\mathrm{D})} \tag{7.107}$$

where $\mu_{\mathrm{C}}^{*,(\mathrm{G})}$ and $\mu_{\mathrm{C}}^{*,(\mathrm{D})}$ are the Gibbs energy values per mole for the graphite phase and the diamond phase, respectively.

After the diamond synthesis reaction has occurred, creating dn moles of diamond, this becomes

$$G + \mathrm{d}G = \left(n_{\mathrm{C}}^{(\mathrm{G})} + \mathrm{d}n_{\mathrm{C}}^{(\mathrm{G})}\right)\mu_{\mathrm{C}}^{*,(\mathrm{G})} + \left(n_{\mathrm{C}}^{(\mathrm{D})} + \mathrm{d}n_{\mathrm{C}}^{(\mathrm{D})}\right)\mu_{\mathrm{C}}^{*,(\mathrm{D})} = G + \left(\mu_{\mathrm{C}}^{*,(\mathrm{D})} - \mu_{\mathrm{C}}^{*,(\mathrm{G})}\right)\mathrm{d}n \tag{7.108}$$

Thus,

$$\mathrm{d}G = \left(\mu_{\mathrm{C}}^{*,(\mathrm{D})} - \mu_{\mathrm{C}}^{*,(\mathrm{G})}\right)\mathrm{d}n \tag{7.109}$$

By considering a closed system which consists of graphite and diamond phases, the criterion for a spontaneous reaction in which graphite undergoes a phase transition, resulting in the creation of diamond, is expressed as $(\mathrm{d}G)_{T,P} < 0$ (Eq. (7.21)).

For the diamond formation reaction, d$n > 0$.

Therefore, based on Eqs. (7.21) and (7.109), the criterion for a spontaneous diamond-formation reaction to take place is

$$\left(\mu_{\mathrm{C}}^{*,(\mathrm{D})} - \mu_{\mathrm{C}}^{*,(\mathrm{G})}\right) < 0 \tag{7.110}$$

Now, we will discuss the conditions for satisfying $\left(\mu_{\mathrm{C}}^{*,(\mathrm{D})} - \mu_{\mathrm{C}}^{*,(\mathrm{G})}\right) < 0$ for diamond formation using the data in Table 7.1. In the interest of simplicity, it is assumed that entropy does not depend on temperature, and density does not depend on pressure.

According to Table 7.1, graphite is stable at a temperature of 298.15 K and pressure of 10^5Pa, as $\mu_{\mathrm{C}}^{*,(\mathrm{G})} < \mu_{\mathrm{C}}^{*,(\mathrm{D})}$. For the following discussion, the chemical potential at 298.15 K, 10^5Pa is written as $\mu_{\mathrm{C},298}^{*,(\mathrm{G})}$ and $\mu_{\mathrm{C},298}^{*,(\mathrm{D})}$ for graphite and diamond, respectively.

Here, by changing the temperature, T, and pressure, P, the condition $\mu_{\mathrm{C}}^{*,(\mathrm{G})} > \mu_{\mathrm{C}}^{*,(\mathrm{D})}$ (which is a function of T and P) is obtained.

Table 7.1 The physicochemical properties of graphite and diamond at 298.15 K and 1 bar

Phase α	$\mu_{\mathrm{C}}^{*,(\alpha)}\ \mathrm{J \cdot mol^{-1}}$	$\overline{S}_{\mathrm{C}}^{*,(\alpha)}\mathrm{J \cdot mol^{-1} \cdot K^{-1}}$	$\overline{V}_{\mathrm{C}}^{*,(\alpha)}\ \mathrm{cm^3 \cdot mol^{-1}}$
Graphite (G)	–1711	5.740	5.310
Diamond (D)	1188	2.377	3.419

According to the basic Eq. (7.28),

$$\Delta\mu_{\mathrm{C}}^{*,(\mathrm{G})} = -\overline{S}_{\mathrm{C}}^{*,(\mathrm{G})}\mathrm{d}T + \overline{V}_{\mathrm{C}}^{*,(\mathrm{G})}\mathrm{d}P \tag{7.111}$$

$$\Delta\mu_{\mathrm{C}}^{*,(\mathrm{D})} = -\overline{S}_{\mathrm{C}}^{*,(\mathrm{D})}\mathrm{d}T + \overline{V}_{\mathrm{C}}^{*,(\mathrm{D})}\mathrm{d}P \tag{7.112}$$

It has been assumed that entropy, $\overline{S}_i^*$, does not depend on temperature and density or, in other words, volume, $\overline{V}_i^*$, does not depend on pressure. Therefore, the change in $\mu_{\mathrm{C}}^{*,(\mathrm{G})}$, i.e. $\Delta\mu_{\mathrm{C}}^{*,(\mathrm{G})}$, and the change in $\mu_{\mathrm{C}}^{*,(\mathrm{D})}$, i.e. $\Delta\mu_{\mathrm{C}}^{*,(\mathrm{D})}$ as a result of the changes in temperature, ΔT, and pressure, ΔP, is

$$\Delta\mu_{\mathrm{C}}^{*,(\mathrm{G})} = -\overline{S}_{\mathrm{C}}^{*,(\mathrm{G})}\Delta T + \overline{V}_{\mathrm{C}}^{*,(\mathrm{G})}\Delta P \tag{7.113}$$

$$\Delta\mu_{\mathrm{C}}^{*,(\mathrm{D})} = -\overline{S}_{\mathrm{C}}^{*,(\mathrm{D})}\Delta T + \overline{V}_{\mathrm{C}}^{*,(\mathrm{D})}\Delta P \tag{7.114}$$

Hence, based on Eq. (7.110), the condition for diamond formation is

$$\mu_{\mathrm{C}}^{*,(\mathrm{G})} = \mu_{\mathrm{C},298}^{*,(\mathrm{G})} - \overline{S}_{\mathrm{C}}^{*,(\mathrm{G})}\Delta T + \overline{V}_{\mathrm{c}}^{*,(\mathrm{G})}\Delta P > \mu_{\mathrm{C},298}^{*,(\mathrm{D})} - \overline{S}_{\mathrm{C}}^{*,(\mathrm{D})}\Delta T + \overline{V}_{\mathrm{C}}^{*,(\mathrm{D})}\Delta P = \mu_{\mathrm{C}}^{*,(\mathrm{D})} \tag{7.115}$$

(1) At constant pressure ($\Delta P = 0$), the condition is

$$\mu_{\mathrm{C},298}^{*,(\mathrm{G})} - \overline{S}_{\mathrm{C}}^{*,(\mathrm{G})}\Delta T > \mu_{\mathrm{C},298}^{*,(\mathrm{D})} - \overline{S}_{\mathrm{C}}^{*,(\mathrm{D})}\Delta T \tag{7.116}$$

Thus,

$$\left(\overline{S}_{\mathrm{C}}^{*,(\mathrm{D})} - \overline{S}_{\mathrm{C}}^{*,(\mathrm{G})}\right)\Delta T > \mu_{\mathrm{C},298}^{*,(\mathrm{D})} - \mu_{\mathrm{C},298}^{*,(\mathrm{G})} \tag{7.117}$$

As can be seen in Table 7.1, $\mu_{\mathrm{C},298}^{*,(\mathrm{D})} - \mu_{\mathrm{C},298}^{*,(\mathrm{G})} = 2899\,\mathrm{J \cdot mol^{-1}}$ and $\overline{S}_{\mathrm{C}}^{*,(\mathrm{D})} - \overline{S}_{\mathrm{C}}^{*,(\mathrm{D})} = -3.363\,\mathrm{J \cdot mol^{-1} \cdot K^{-1}}$.

Hence,

$$\Delta T > \frac{\mu_{\mathrm{C},298}^{*,(\mathrm{D})} - \mu_{\mathrm{C},298}^{*,(\mathrm{G})}}{\overline{S}_{\mathrm{C}}^{*,(\mathrm{D})} - \overline{S}_{\mathrm{C}}^{*,(\mathrm{G})}} = -862\,\mathrm{K} \tag{7.118}$$

$$\Delta T = T - 298.15 \tag{7.119}$$

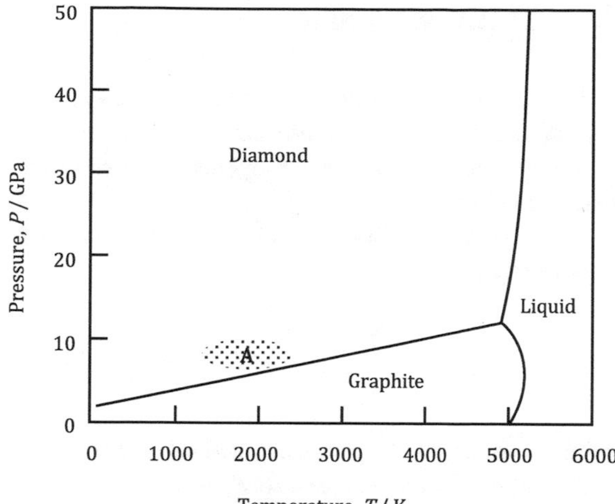

Fig. 7.2 Pressure-temperature diagram for carbon (adapted from F. P. Bundy et al., Carbon, 34 (1996) 141–153.)

Therefore, it must be the case that $T < -564$ K (which is, of course, impossible) for a diamond phase to form. This result implies that a stable diamond phase cannot be synthesised solely by changing the temperature.

(2) At constant temperature ($\Delta T = 0$), the condition is

$$\mu_{\mathrm{C},298}^{*,(\mathrm{G})} + \overline{V}_{\mathrm{C}}^{*,(\mathrm{G})} \Delta P > \mu_{\mathrm{C},298}^{*,(\mathrm{D})} + \overline{V}_{\mathrm{C}}^{*,(\mathrm{D})} \Delta P \tag{7.120}$$

Thus,

$$-\left(\overline{V}_{\mathrm{C}}^{*,(\mathrm{D})} - \overline{V}_{\mathrm{C}}^{*,(\mathrm{G})}\right) \Delta P > \mu_{\mathrm{C},298}^{*,(\mathrm{D})} - \mu_{\mathrm{C},298}^{*,(\mathrm{G})} \tag{7.121}$$

As can be seen in Table 7.1, $\mu_{\mathrm{C},298}^{*,(\mathrm{D})} - \mu_{\mathrm{C},298}^{*,(\mathrm{G})} = 2899\ \mathrm{J \cdot mol^{-1}}$ and $\overline{V}_{\mathrm{C}}^{*,(\mathrm{D})} - \overline{V}_{\mathrm{C}}^{*,(\mathrm{G})} = -1.891 \times 10^{-6}\mathrm{m^3 \cdot mol^{-1}}$.

Hence,

$$\Delta P > -\frac{\mu_{\mathrm{C},298}^{*,(\mathrm{D})} - \mu_{\mathrm{C},298}^{*,(\mathrm{G})}}{\overline{V}_{\mathrm{C}}^{*,(\mathrm{D})} - \overline{V}_{\mathrm{C}}^{*,(\mathrm{G})}} = 1.53 \times 10^6\ \mathrm{J \cdot m^{-3}} \approx 1.5 \times 10^9\ \mathrm{Pa} \tag{7.122}$$

$$\Delta P = P - 10^5\ \mathrm{Pa} \tag{7.123}$$

Therefore, if the pressure, P, is greater than 1.5×10^9 Pa at 298.15 K, it is thermodynamically possible to synthesise a stable diamond phase. In a real-world process, diamonds are synthesised at higher than 1273 K (1000 °C) and at approximately 10^{10} Pa (Region A in Fig. 7.2).

7.4 Chemical Equilibrium

In this section, we derive the conditions for chemical equilibrium by applying the general conditions for equilibrium to a system that involves chemical reactions.

Firstly, let us discuss a homogeneous closed system in which a single chemical reaction occurs.

The general equation for a chemical reaction with r chemical substances may be expressed as

$$\sum_{i=1}^{r} \nu_{\mathrm{i}} \mathrm{X}_{\mathrm{i}} = 0 \tag{7.124}$$

where X_i is 1 mol of a pure chemical substance in a stable state of aggregation at the temperature, T, and the pressure, P, at which the reaction occurs, and ν_i is the number of moles of component i formed by the reaction. Note that in the other chapters and sections of this book, this reaction has been written as

$$\nu_1' \mathrm{X}_1 + \nu_2' \mathrm{X}_2 = \nu_3' \mathrm{X}_3 + \nu_4' \mathrm{X}_4 \tag{7.125}$$

and the sign of ν_1', ν_2', ν_3', and ν_4' is positive. However, in Eq. (7.124), the ν_{i} values for reactants are positive, whereas those of products are negative.

For example, the reaction

$$\mathrm{H_2} + \frac{1}{2}\mathrm{O_2} = \mathrm{H_2O} \tag{7.126}$$

may be rewritten as

$$\mathrm{H_2O} - \mathrm{H_2} - \frac{1}{2}\mathrm{O_2} = 0 \tag{7.127}$$

In this case, $\nu_1 = -1$, $\nu_2 = -\frac{1}{2}$, and $\nu_3 = +1$. $X_1 = H_2$, $X_2 = O_2$, and $X_3 = H_2O$ according to Eq. (7.124).

The differential increments (dn_i) of the number of moles of component i ($i = 1, \ldots\ldots, r$), produced in the reaction described in Eq. (7.124), are related as follows, and can also be related to the *extent of reaction* (also known as *progress variable* or *reaction coordinate*), ξ:

$$\frac{dn_1}{\nu_1} = \frac{dn_2}{\nu_2} = \ldots\ldots = \frac{dn_r}{\nu_r} = d\xi \tag{7.128}$$

$d\xi$ is assumed to be uniform throughout a one-phase system. If it is not uniform, composition gradients will be formed and diffusion fluxes will be generated in turn. Such a condition featuring diffusion fluxes differs from that described above for a closed homogeneous system that results only from chemical reactions.

As shown in Eq. (7.20), a criterion for equilibrium for a closed system at constant temperature, T, and constant pressure, P, is

$$(\mathrm{d}G)_{T,P} \geq 0 \tag{7.129}$$

As shown in Eq. (7.57), for a closed system in which the masses of components are changed as a result of a reaction, $\mathrm{d}G$ is

$$\mathrm{d}G = -S\mathrm{d}T + V\mathrm{d}P + \sum_{i=1}^{r} \mu_i \mathrm{d}n_i \tag{7.130}$$

where S is the entropy, T is the temperature, V is the volume, P is the pressure, μ_i is the chemical potential of component i, and n_i is the number of moles of component i.

Hence, the variation in Gibbs energy at constant T and P can be expressed as

$$(\mathrm{d}G)_{T,P} = \sum_{i=1}^{r} \mu_i \mathrm{d}n_i \tag{7.131}$$

$\mathrm{d}n_i$ is expressed using ξ as follows:

$$\mathrm{d}n_i = \nu_i \mathrm{d}\xi \tag{7.132}$$

Therefore,

$$(\mathrm{d}G)_{T,P} = \left(\sum_{i=1}^{r} \mu_i \nu_i\right) \mathrm{d}\xi \tag{7.133}$$

Hence, the criterion for equilibrium for a system that involves chemical reactions is, as shown in Eq. (7.20),

$$(\mathrm{d}G)_{T,P} = \left(\sum_{i=1}^{r} \mu_i \nu_i\right) \mathrm{d}\xi \geq 0 \tag{7.134}$$

This equation must be true for all conceivable variations of ξ. Therefore, if $\mathrm{d}\xi > 0$, $\sum_{i=1}^{r} \mu_i \nu_i \geq 0$ is a criterion for equilibrium; if $\mathrm{d}\xi < 0$, $\sum_{i=1}^{r} \mu_i \nu_i \leq 0$ is a criterion.

$\mathrm{d}\xi$ can be either positive or negative when all r substances are present in a reaction mixture. Hence, Eqs. (7.133) and (7.134) are simultaneously valid when

$$\sum_{i=1}^{r} \mu_i \nu_i = 0 \tag{7.135}$$

This is the condition for the equilibrium of a system containing r chemically reactive substances.

Now, we will derive the criterion for equilibrium in terms of internal energy, U, using Eq. (7.9). This can be performed in a similar manner to the derivation of Eq. (7.135):

As shown in Eq. (7.9), a criterion for equilibrium for a closed system at constant entropy, S, and constant volume, V, is

$$(\mathrm{d}U)_{S,V} \geq 0 \tag{7.136}$$

As shown in Eq. (7.49), for a closed system in which the individual masses of components are changed as a result of a reaction, dU is

$$\mathrm{d}U = T\mathrm{d}S - P\mathrm{d}V + \sum_{i=1}^{r} \mu_i \mathrm{d}n_i \tag{7.137}$$

Hence, the variation in internal energy at constant S and V ($\mathrm{d}S = 0$, $\mathrm{d}V = 0$) can be expressed by

$$(\mathrm{d}U)_{T,P} = \sum_{i=1}^{r} \mu_i \mathrm{d}n_i \tag{7.138}$$

and $\mathrm{d}n_i$ is expressed by ξ as follows:

$$\mathrm{d}n_i = \nu_i \mathrm{d}\xi \tag{7.139}$$

Therefore,

$$(\mathrm{d}U)_{S,V} = \left(\sum_{i=1}^{r} \mu_i \nu_i \right) \mathrm{d}\xi \geq 0 \tag{7.140}$$

is a criterion for equilibrium.

As was discussed above, $\mathrm{d}\xi$ can be either positive or negative. Hence, the condition for equilibrium for a system containing r chemically reactive substances is

$$\sum_{i=1}^{r} \mu_i \nu_i = 0 \tag{7.141}$$

(End of the derivation)

For a closed homogeneous system in which several (q) reactions occur, the following condition for equilibrium can be obtained:

$$\sum_{i=1}^{r} \mu_i \nu_i^{\sigma} = 0, \quad \sigma = 1, \ldots\ldots, q \tag{7.142}$$

For a closed heterogeneous system in which a single chemical reaction occurs and both reactants and products exist in at least one of the phases, the following condition for equilibrium can be obtained:

$$\sum_{i=1}^{r} \mu_i \nu_i = 0 \tag{7.143}$$

The derivation of Eqs. (7.142) and (7.143) is omitted due to space constraints, but a complete discussion may be found in Kirkwood and Oppenheim (1961, p. 102–103).

7.5 The Phase Rule

The phase rule fixes the number of independent intensive variables that is required to determine the thermodynamic state of a heterogeneous system at equilibrium.

What is the number of intensive variables such as T, P, and concentration that we can freely determine for a heterogeneous system at equilibrium? How many intensive variables are required to determine the state of the system? The phase rule derived by Gibbs provides answers to these questions, and plays an important role in creating e.g. phase diagrams.

7.5.1 Systems in Which no Chemical Reaction Occurs

In this section, we consider the phase rule for a system in which no chemical reaction is occurring using the general criteria for equilibrium for a heterogeneous system derived in Sect. 7.3. The equilibrium state of a heterogeneous system as discussed in this chapter does not consider interactions with phases that exist outside of the system; if these occur, the phases must be considered as part of the system. The heterogeneous system considered in this chapter is an isolated heterogeneous system, as discussed in Sect. 7.3.

Let us recall the assumptions that we made to derive the conditions that intensive properties must satisfy in order for a heterogeneous system to establish an equilibrium: (1) The influence of the surface is negligible, and so the extensive properties, such as U, S, and V, are expressed as summations of each contribution from the bulk phase. (2) The only external force acting on the system is a uniform normal pressure. (3) All interfaces are deformable, heat conductive, and pervious to all of the components. (4) No chemical reactions occur.

In an equilibrium state, a system that consists of ν phases and r independent components is, according to Eqs. (7.86)–(7.88) restricted by the following conditions:

$$\begin{array}{ccccccccc} & 1 & & 2 & & 3 & & & \nu-1 \\ 1 & P^{(1)} & = & P^{(2)} & = & P^{(3)} & = & \cdots\cdots\cdots\cdots = & P^{(\nu)} \\ 2 & T^{(1)} & = & T^{(2)} & = & T^{(3)} & = & \cdots\cdots\cdots\cdots = & T^{(\nu)} \\ 1+2 & \mu_1^{(1)} & = & \mu_1^{(2)} & = & \mu_1^{(2)} & = & \cdots\cdots\cdots\cdots = & \mu_1^{(\nu)} \\ & \cdot & & \cdot & & \cdot & & & \cdot \\ & \cdot & & \cdot & & \cdot & & & \cdot \\ & \cdot & & \cdot & & \cdot & & & \cdot \\ & \cdot & & \cdot & & \cdot & & & \cdot \\ & \cdot & & \cdot & & \cdot & & & \cdot \\ r+2 & \mu_r^{(1)} & = & \mu_r^{(2)} & = & \mu_r^{(2)} & = & \cdots\cdots\cdots\cdots = & \mu_r^{(\nu)} \end{array} \tag{7.144}$$

Therefore, there are $(r+2)(\nu-1)$ conditions for this system. As is discussed above (Sect. 2.3), the intensive properties of phase α are determined by $r+1$ variables, i.e. $T^{(\alpha)}, P^{(\alpha)}, x_1^{(\alpha)}, x_2^{(\alpha)} \ldots\ldots x_{r-1}^{(\alpha)} (\alpha = 1, 2, \ldots\ldots \nu)$ if the only external force exerted on the system is a uniform normal pressure.

The number of variables that define the equilibrium state of the ν isolated phases is thus $\nu(r+1)$. The number of variables, f, of the set $T^{(\alpha)}, P^{(\alpha)}, x_1^{(\alpha)}, x_2^{(\alpha)} \ldots\ldots x_{r-1}^{(\alpha)} (\alpha = 1, 2, \ldots\ldots \nu)$ that can be specified within the field of heterogeneous equilibrium states is therefore

$$f = \nu(r+1) - (r+2)(\nu-1) = r + 2 - \nu \tag{7.145}$$

This equation may be rewritten as follows using the symbols recommended by the IUPAC:

$$\mathcal{F} = \mathcal{C} - \mathcal{P} + 2 \tag{7.146}$$

where $\mathcal{F}$ is the number of degrees of freedom of the system, $\mathcal{C}$ is the number of components (chemical species) in the system, and $\mathcal{P}$ is the number of phases in the system.

Equation (7.146) is known as the *Gibbs' phase rule*.

In equilibrium phase diagrams for e.g. alloys, pressure, P, is implicitly defined as $P = 10^5$ Pa. Therefore, the number of degrees of freedom is decreased by 1 and the equation becomes

$$\mathcal{F} = \mathcal{C} - \mathcal{P} + 1 \tag{7.147}$$

7.5.2 Systems in Which Chemical Reactions Occur

By estimating the number of independent components, i, in a system in which chemical reactions take place, the form shown in Eq. (7.144) is true and thus Eq. (7.145) ($f = r + 2 - \nu$) can be applied. However, and as we see below, the number of chemical species in the system is normally used for the number of components when calculating the phase rule. If we use this counting method to count the number of components, further restraints are imposed on the phase rule, i.e. the condition for equilibrium derived in Sect. 7.4 $\left(\sum_{i=1}^{r} \nu_i \mu_i = 0\right)$ must be taken into account for each independent reaction, and thus the number of degrees of freedom is decreased by the number of independent reactions, q.

As shown in Example 7.4, if either products or reactants exist in a homogeneous system before a reaction, the amount of reactant that is produced in the reaction as it progresses towards the equilibrium position is determined only by the chemical reaction equation. Therefore, the restrictions on the concentration that are defined by the number of reactions, s, are imposed on the phase rule. In this case, the number of degrees of freedom, $\mathcal{F}$, is thus

$$\mathcal{F} = \mathcal{C} - \mathcal{P} + 2 - (q + s) \tag{7.148}$$

where $\mathcal{C}$ is the number of all stable chemical species in the system. 'All stable' means that when components are produced by the reactions of components in a system, they are also counted as 'apparent' independent components.

The components produced by the reactions are, in essence, not independent components. The number of independent components, $\mathcal{C}'$, is expressed as

$$\mathcal{C}' = \mathcal{C} - q \tag{7.149}$$

Example 7.3

For the phase diagram shown in Fig. 7.3, calculate the number of degrees of freedom of the system, in which (1) ℓ and α phases coexist and, (2) ℓ, α, and β phases coexist.

Solutions:

(1)

$$\mathcal{C} = 2(1 \text{ and } 2)$$
$$\mathcal{P} = 2(\ell \text{ and } \alpha)$$
$$s = q = 0$$

Hence,

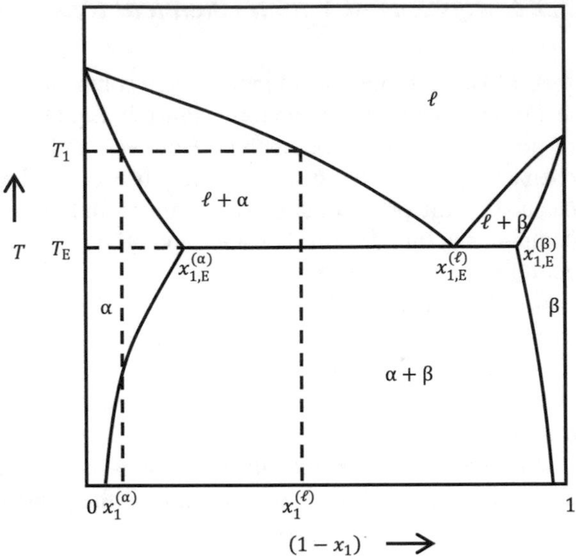

Fig. 7.3 Phase diagram of a binary system

$$\mathcal{F} = \mathcal{C} - \mathcal{P} + 2 - (q + s) = 2$$

The number of degrees of freedom is 1 at $P = 10^5$ Pa (constant pressure). Therefore, if the temperature of the system is specified as T_1, the concentrations of the liquid phase and α phase are determined as $x_1^{(\ell)}$ and $x_1^{(\alpha)}$, respectively, as shown in Fig. 7.3.

(2)

$$\mathcal{C} = 2$$
$$\mathcal{P} = 3(\ell,\ \alpha,\ \text{and}\beta)$$
$$s = q = 0$$

Hence,

$$\mathcal{F} = \mathcal{C} - \mathcal{P} + 2 - (q + s) = 1$$

The number of degrees of freedom is zero at $P = 10^5$ Pa (constant pressure). Therefore, for a system in which ℓ, α, and β phases coexist, the temperature of the system and the concentration of each phase ($x_{1,E}^{(\ell)}$, $x_{1,E}^{(\alpha)}$, and $x_{1,E}^{(\beta)}$) are determined from the phase diagram.

Example 7.4

Calculate the number of degrees of freedom of the following reaction system:

$$2H_2O(g) = 2H_2(g) + O_2(g)$$

Solution:

$$\mathcal{C} = 3(H_2O, H_2, \text{and } O_2)$$
$$\mathcal{P} = 1(\text{gas phase})$$
$$q = 1$$

Regarding the number of reactions, s:

(1) If only H_2O exists before the reaction, i.e. the reaction that occurs is the decomposition of water, the mole ratio of $H_2(g)$ to $O_2(g)$ is 2:1. Thus, $s = 1$.

Thus,

$$\mathcal{F} = \mathcal{C} - \mathcal{P} + 2 - (q + s) = 2$$

If the temperature, T, and total pressure, P, are specified, from the equilibrium constant, K

$$K = \frac{P_{H_2}^2 P_{O_2}}{P_{H_2O}^2}$$
$$\frac{P_{H_2}}{P_{O_2}} = 2$$

and

$$P_{H_2O} + P_{H_2} + P_{O_2} = P$$

K and P are known. Therefore, P_{H_2O}, P_{H_2}, and P_{O_2} can be determined. (see Sects. 10.1 and 10.2 for the details of the equilibrium constant, K)

(2) If H_2 and O_2 also exist in addition to H_2O before the reaction, $s = 0$.

Hence,

$$\mathcal{F} = \mathcal{C} - \mathcal{P} + 2 - (q + s) = 3$$

In this case, another intensive variable, aside from T and P (e.g. partial pressure of H_2O, H_2, or O_2), is required in order to determine the partial pressure of the other components.

Example 7.5

Calculate the number of degrees of freedom of the following decomposition reaction:

$$CaCO_3(s) = CaO(s) + CO_2(g)$$

Solution:

$$\mathcal{C} = 3(CaCO_3, CaO, \text{and } CO_2)$$
$$\mathcal{P} = 3(CaCO_3(s), CaO(s), \text{and } CO_2(g))$$
$$q = 1$$

CaO and CO_2 are produced by the reaction, and the mole ratio of CaO to CO is 1:1.

However, these are separated into solid and gas phases. Therefore, the gas phase consists of $CO_2(g)$ produced by the decomposition reaction and CaO(g) (equilibrium vapour pressure) as determined by the equilibrium with CaO(s), and so there is no restriction on the concentration in relation to the gas phase. Hence, $s = 0$.

Thus,

$$\mathcal{F} = \mathcal{C} - \mathcal{P} + 2 - (q + s) = 1$$

If a temperature is specified, P_{CO_2} can be determined.

The following reactions can also be considered:

$$CaO(s) = CaO(g)$$
$$CaCO_3(s) = CaCO_3(g)$$

However, as any limitations imposed as a result of these reactions are already considered (counted) by the condition for heterogeneous equilibrium, $\mu_i^{(\alpha)} = \mu_i^{(\beta)}$, no additional limitation is required to calculate the number of degrees of freedom.

Example 7.6

Calculate the number of degrees of freedom of a system consisting of a silica (SiO_2) crucible containing molten iron (Fe, Si, O), slag (SiO_2, FeO), and gas (O_2).

Solution:

$$\mathcal{C} = 6(SiO_2, Fe, Si, O, FeO, \text{and } O_2)$$
$$\mathcal{P} = 4(SiO_2(s), \text{molten iron, slag, and } O_2(g))$$

In order to obtain q, the number of individual chemical reactions must be counted.
Three reactions are occurring, therefore $q = 3$.

$$\underline{Si} + 2\underline{O} = SiO_2(s) \quad (1)$$
$$2\underline{O} = O_2(g) \quad (2)$$
$$Fe(l) + \underline{O} = (FeO) \quad (3)$$

Other reactions, such as $\underline{Si} + O_2(g) = SiO_2(s)$, can be obtained by combining Reactions (1) and (2).
It should be noted that $s = 0$.
Thus,

$$\mathcal{F} = \mathcal{C} - \mathcal{P} + 2 - (q + s) = 1$$

As shown in this example, for a complex heterogeneous system, a special effort must be made to count the number of components, r, and the number of individual chemical reactions, q. As can be seen in the derivation of the phase rule, all of the components exist in all of the phases. Therefore, it is possible to obtain the number of components by counting the number of components in one of the phases.

Example 7.7
Consider a system in which the influence of an interface is not negligible.
For example, consider the number of degrees of freedom of a system in which a water droplet (the radius of which is r) and water vapour are in equilibrium.
Solution:
If the droplet is sufficiently large and the surface contribution is negligible, the already-derived phase rule, Eq. (7.145), can be applied:

$$\mathcal{C} = 1(H_2O)$$
$$\mathcal{P} = 2(H_2O(\ell),\ H_2O(g), \text{and } O_2)$$

$q = s = 0$
Thus,

$$\mathcal{F} = \mathcal{C} - \mathcal{P} + 2 - (q + s) = 1$$

However, when the radius r is sufficiently small the vapour pressure changes, as is shown in Eq. (7.150) (the *Kelvin equation*).

$$p^{(\alpha)} - p^{(\beta)} = \frac{2\sigma}{r} \quad (7.150)$$

where P is the vapour pressure at radius ∞, P_r is the vapour pressure at radius r , σ is the surface tension of the droplet, and $\overline{V}^{(\ell)}$ is the molar volume of the droplet.

When the radius of the droplet is sufficiently small and the specific surface area is sufficiently large, the contribution of the surface is not negligible and, as such, the vapour pressure is affected by T and r . Thus, the number of degrees of freedom is 2 and cannot be explained by Eq. (7.146) as this equation does not factor in the contribution of the surface. The phase rule, which considers the contribution of a surface, is

$$\mathcal{F} = 1 + (\mathcal{C} - q) \quad (7.151)$$

$\mathcal{C} = 1$ and $q = 0$, thus $\mathcal{F} = 2$. The example of the small droplet can thus be explained using this equation.

A complete discussion of the phase rule that considers the contribution of a surface can be found in Defay and Prigogine (1966).

Example 7.8

Calculate the number of degrees of freedom of a $CaO(s)-CaCO_3(s)-CO_2(g)-Ar(g)$ system.

Solution:

In this system, the following reaction occurs:

$CaCO_3(s) = CaO(s) + CO_2(g)$

$$\mathcal{C} = 4(CaCO_3, CaO, CO_2, \text{and } Ar)$$

$$\mathcal{P} = 3(CaO(s), CaCO_3(s), \text{and gas phase})$$

$q = 1$

$s = 0$

Hence,

$$\mathcal{F} = \mathcal{C} - \mathcal{P} + 2 - (q + s) = 2$$

Example 7.9

Calculate the number of degrees of freedom of a $Fe(s) - Fe_3O_4(s) - H_2(g) - H_2O(g)$ system.

Solution:

In this system, the following reaction occurs:

$$Fe_3O_4(s) + 4H_2(g) = 3Fe(s) + 4H_2O(g)$$

$$\mathcal{C} = 4(Fe_3O_4, H_2, Fe, \text{and } H_2O)$$
$$\mathcal{P} = 3(Fe_3O_4(s), Fe(s), \text{and gas phase})$$

$q = 1$
$s = 0$
Hence,

$$\mathcal{F} = \mathcal{C} - \mathcal{P} + 2 - (q + s) = 2$$

Example 7.10

Calculate the number of degrees of freedom of a $SiC(s) - C(s) -$ molten iron (Fe, Si, C) system.

Solution:

In this system, the following reaction occurs:

$$C(s) + \underline{Si} = SiC(s)$$

$$\mathcal{C} = 4(SiC, C, Fe, \text{and } Si)$$
$$\mathcal{P} = 3(SiC\,(s), C(s), \text{and molten iron})$$

$q = 1$
$s = 0$
Hence,

$$\mathcal{F} = \mathcal{C} - \mathcal{P} + 2 - (q + s) = 2$$

Example 7.11

Calculate the number of degrees of freedom of a $Si_3N_4(s)$ – molten iron (Fe, Si, N) – $N_2(g)$ – Ar(g) system.

Solution:

In this system, the following reactions occur:

$$\frac{1}{2}N_2(g) = \underline{N}$$

$$4\underline{N} + 3Si = Si_3N_4(s)$$

$$\mathcal{C} = 6(Si_3N_4, \text{Fe, Si, N}, N_2, \text{and Ar})$$

$$\mathcal{P} = 3(Si_3N_4(s), \text{molten iron, and gas phase})$$

$q = 2$

$s = 0$

Hence,

$$\mathcal{F} = \mathcal{C} - \mathcal{P} + 2 - (q + s) = 3$$

Example 7.12

Calculate the number of degrees of freedom of the following reaction (Si deoxidation equilibrium in liquid iron):

$$\underline{Si} + 2\underline{O} = SiO_2(s)$$

Solution:

$$\mathcal{C} = 4(SiO_2, \text{Fe, Si, and O})$$

$$\mathcal{P} = 2(SiO_2(s) \text{ and molten iron})$$

$q = 1$

$s = 0$

Hence,

$$\mathcal{F} = \mathcal{C} - \mathcal{P} + 2 - (q + s) = 3$$

7.6 Summary

The criterion for equilibrium:
The criterion for equilibrium for a closed system is

$$dU + PdV - TdS \geq 0 \quad (7.6)$$

where U is the internal energy, P is the pressure, V is the volume, T is the temperature, and S is the entropy.

The criteria for equilibrium under more restricted conditions can be obtained based on the above general criteria (Eqs. (7.7), (7.9), (7.10), and (7.20)).

Helmholtz energy and Gibbs energy:
Helmholtz energy, F, is determined as follows:

$$F = U - TS \quad (7.11)$$

where U is the internal energy, T is the temperature, and S is the entropy.

Gibbs energy, G, is determined as follows:

$$G = U + PV - TS = H - TS \quad (7.17)$$

where U is the internal energy, P is the pressure, V is the volume, T is the temperature, S is the entropy, and H is the enthalpy.

Maxwell's relations:
Maxwell's relations Eqs. (7.40)–(7.43), which allow non-measurable quantities to be calculated, can be deduced from the basic equations for equilibrium.

Chemical potential:
Chemical potential is a partial molar quantity of internal energy, U, enthalpy, H, Helmholtz energy, F, and Gibbs energy, G, under specific conditions Eqs. (7.60)–(7.63).

Gibbs-Duhem equation:
The Gibbs-Duhem equation allows us to calculate e.g. activity and molar volume.

$$\sum_{i=1}^{r} x_i d\mu_i + \overline{S}dT - \overline{V}dP = 0 \quad (7.67)$$

where x_i is the mole fraction of component i, μ_i is the chemical potential of component H, $\overline{S}$ is the mean molar entropy, T is the temperature, $\overline{V}$ is the mean molar volume, and P is the pressure.

At constant T and P, the equation reduces to

$$\sum_{i=1}^{r} n_i \mathrm{d}\mu_i = \sum_{i=1}^{r} x_i \mathrm{d}\mu_i = 0 \tag{7.68}$$

where n_i is the number of moles of component i.

Conditions for equilibrium:
The necessary conditions for equilibrium when all interfaces are (1) deformable, (2) heat conductors, and (3) pervious to all of the components are

$$P^{(\alpha)} = P \text{ for all } \alpha \tag{7.86}$$

$$T^{(\alpha)} = T \text{ for all } \alpha \tag{7.87}$$

$$\mu_i^{(\alpha)} = \mu_i \text{ for all } \alpha \text{ and } i \, (i = 1, \ldots\ldots, r) \tag{7.88}$$

Clausius-Clapeyron equation:
The influence of temperature and pressure on the phase transformation of solid, melting, and evaporation phenomena can be described by the Clausius-Clapeyron equation:

$$\frac{\mathrm{d}T}{\mathrm{d}P} = \frac{T \Delta \overline{V}^*}{\Delta \overline{H}^*} \tag{7.104}$$

where T and P are the temperature and the pressure, respectively; when equilibrium has been established between Phases 1 and 2, $\Delta \overline{V}^* = \overline{V}^{*,(2)} - \overline{V}^{*,(1)}$ is the molar volume change accompanying the phase transformation and $\Delta \overline{H}^* = \overline{H}^{*,(2)} - \overline{H}^{*,(1)}$ is the phase transition heat per mole (molar latent heat).

Phase rule:
The number of degrees of freedom, $\mathcal{F}$, is expressed by

$$\mathcal{F} = \mathcal{C} - \mathcal{P} + 2 - (q + s) \quad \text{(phase rule)}$$

where $\mathcal{C}$ is the number of stable chemical species in the system, $\mathcal{P}$ is the number of phases in the system, q is the number of independent reactions, and s is the number of reactions.

The condition for equilibrium:
The condition for equilibrium of a system that contains r chemically reactive substances is

$$\sum_{i=1}^{r} \mu_i \nu_i = 0 \tag{7.135}$$

where μ_i is the chemical potential of component i and ν_i is the number of moles of component i formed by the reaction.

References

Defay R, Prigogine I (1966) Surface tension and adsorption. Longmans, Green and Co Ltd.
Kirkwood JG and Oppenheim I (1961) Chemical thermodynamics, 1st edn. McGraw-Hill

Chapter 8
Chemical Potential and Activity

In this chapter, the chemical potential (i.e. Gibbs energy per mole) of several different systems is discussed together with activity (Sects. 8.1 and 8.2). Several standard states that are commonly used to describe activity (Raoultian standard state, Henrian standard state, etc.) are explained (Sect. 8.2) and methods of measuring and calculating activity are discussed (Sects. 8.3 and 8.4).

8.1 Chemical Potential and the Concentration of an Ideal Mixture

We have discussed the condition of the equilibrium in relation to mass for heterogeneous systems Eq. (7.88) and chemical equilibrium Eq. (7.143) using chemical potential. However, in practice, it is useful to know the chemical potential at equilibrium as a function of concentration or partial pressure. It is no exaggeration to state that most discussions in the field of thermodynamics post-J. W. Gibbs relate chemical potential to concentration.

Let us consider ideal mixtures in order to clarify the relationship between chemical potential and concentration. It is common in thermodynamics to start a discussion with the ideal case, and expand to consider non-ideal cases by comparing the results, i.e. assessing the level of deviation from the ideal case.

T. Matsushita and K. Mukai, *Chemical Thermodynamics in Materials Science*,
https://doi.org/10.1007/978-981-13-0405-7_8

8.1.1 Ideal Gases

8.1.1.1 The Relationship Between Chemical Potential and Pressure for Pure Substances

According to Eq. (7.32),

$$\left(\frac{\partial G}{\partial P}\right)_T = V \tag{8.1}$$

where G is the Gibbs energy, P is the pressure, T is the temperature, and V is the volume.

Therefore, the following relationship is true for an ideal gas that consists of 1 mol of pure component i:

$$\begin{aligned}\left(\frac{\partial \mu_i^{*,\mathrm{id(g)}}}{\partial P}\right)_T &= \overline{V}_i^{*,\mathrm{id(g)}} \\ &= \frac{RT}{P}\end{aligned} \tag{8.2}$$

where $\mu_i^{*,\mathrm{id(g)}}$ is the chemical potential of pure component i in an ideal gas phase (the superscripts $*$, id, and (g) denote pure, ideal, and gas phase, respectively), $\overline{V}_i^{*,\mathrm{id(g)}}$ is the molar volume of pure component i, T is the temperature, and R is the gas constant $\left(= 8.314\ \mathrm{J \cdot mol^{-1} \cdot K^{-1}}\right)$.

By integrating the equation from P° (pressure at standard state) to P, we obtain

$$\mu_i^{*,\mathrm{id(g)}}(P) - \mu_i^{\mathrm{o,id(g)}}(P^\circ) = RT\ \ln(P/P^\circ) \tag{8.3}$$

where $\mu_i^{*,\mathrm{id(g)}}(P)$ is the chemical potential of component i at pressure P and $\mu_i^{\mathrm{o,id(g)}}(P^\circ)$ is the chemical potential of component i in a standard state ($P = P^\circ$) for an ideal gas. The superscript o denotes a standard state. The pressure at standard state, P°, is 1 bar = 10^5 Pa.[1]

For the sake of simplicity, we can use bar as a unit of pressure instead of Pa, reducing the equation to

$$\mu_i^{*,\mathrm{id(g)}} - \mu_i^{\mathrm{o,id(g)}} = RT\ \ln P \tag{8.4}$$

Hence, the chemical potential of pure component i in an ideal gas phase can be described as

[1] 1 atm was widely used as the standard pressure until 1982, but today 1 bar $\left(= 10^5\ \mathrm{Pa}\right)$ is used. In practice, the pressure difference between 1 atm and 1 bar is generally considered to be negligible, as 1 atm = 1.01325 bar and this difference is negligible in comparison to the uncertainty of the available data.

$$\mu_i^{*,\mathrm{id(g)}} = \mu_i^{\mathrm{o,id(g)}} + RT \ln P \tag{8.5}$$

Here, there is a linear relationship between the chemical potential of the pure component i of ideal gas phase and ln P at a constant temperature.

8.1.1.2 The Relationship Between Chemical Potential and Partial Pressure or Concentration for Gas Mixtures

We next consider the mixture of an ideal gas. When ideal gases consisting of n_1 moles of Component 1 and n_2 moles of Component 2 are mixed at constant temperature T and constant pressure P, the *Gibbs energy of mixing* (i.e. the change in Gibbs energy as a result of the mixing—see Chap. 9 for an in-depth discussion of this), $\Delta_{\mathrm{mix}}G$, is

$$\Delta_{\mathrm{mix}}G = \underbrace{n_1\mu_1 + n_2\mu_2}_{\text{After the mixing}} - \underbrace{\left(n_1\mu_1^* + n_2\mu_2^*\right)}_{\text{Before the mixing}} \tag{8.6}$$

where μ_1 and μ_2 are the chemical potentials of Components 1 and 2 in the mixture, respectively, and μ_1^* and μ_2^* are the chemical potentials of pure Components 1 and 2, respectively. Here, μ_1^* and μ_2^* are functions of the pressure, P.

As Gibbs energy, G, after mixing is

$$G = n_1\mu_1 + n_2\mu_2 \tag{8.7}$$

the Gibbs energy of the mixture, which consists of n_1 moles of Component 1 and n_2 moles of Component 2, is

$$\begin{aligned} G &= n_1\mu_1 + n_2\mu_2 \\ &= \Delta_{\mathrm{mix}}G + n_1\mu_1^* + n_2\mu_2^* \end{aligned} \tag{8.8}$$

$\Delta_{\mathrm{mix}}G$ can also be described as follows:

$$\Delta_{\mathrm{mix}}G = RT(n_1 \ln x_1 + n_2 \ln x_2) \tag{8.9}$$

where R is the gas constant, T is the temperature, and x_i is the mole fraction of component i.

Equation (8.9) can be derived as follows:

As shown in Eq. (7.18),

$$\mathrm{d}G = \mathrm{d}U + P\mathrm{d}V + V\mathrm{d}P - T\mathrm{d}S - S\mathrm{d}T \tag{8.10}$$

where U is the internal energy, P is the pressure, V is the volume, T is the temperature, and S is the entropy.

Therefore, at constant T and P (i.e. $\Delta T = 0,\ \Delta P = 0$), $\Delta_{\mathrm{mix}}G$ is

$$\Delta_{\text{mix}}G = \Delta_{\text{mix}}U + P\Delta_{\text{mix}}V - T\Delta_{\text{mix}}S \tag{8.11}$$

where Δ_{mix} denotes the change in each property as a result of the mixing. $\Delta_{\text{mix}}U$, $\Delta_{\text{mix}}V$, and $\Delta_{\text{mix}}S$ are the *internal energy change of mixing*, the *volume change of mixing*, and the *entropy change of mixing*, respectively (i.e. the change in internal energy, volume, and entropy as a result of mixing, respectively).

As *Dalton's law* is true for an ideal gas, $\Delta_{\text{mix}}V = 0$. This means that, for an ideal gas mixture, the total pressure exerted is equal to the sum of the pressures exerted by each of the individual component gases. In other words, the total volume is constant for mixing at constant T and P. In addition, there is no interaction between the molecules and atoms for ideal gases, and so $\Delta_{\text{mix}}U = 0$.

According to statistical thermodynamics, the entropy of mixing, $\Delta_{\text{mix}}S$, is described as follows Eq. (6.31):

$$\Delta_{\text{mix}}S = -R(n_1 \ln x_1 + n_2 \ln x_2) \tag{8.12}$$

Hence, using Eq. (8.11), we obtain

$$\Delta_{\text{mix}}G = RT(n_1 \ln x_1 + n_2 \ln x_2) \tag{8.13}$$

(End of the derivation of Eq. (8.9))

By substituting Eq. (8.9) into Eq. (8.8), we obtain

$$\begin{aligned} G &= \Delta_{\text{mix}}G + n_1\mu_1^* + n_2\mu_2^* \\ &= RT(n_1 \ln x_1 + n_2 \ln x_2) + n_1\left(\mu_1^{\text{o,id(g)}} + RT\ \ln P\right) \\ &\quad + n_2\left(\mu_2^{\text{o,id(g)}} + RT\ \ln P\right) \end{aligned} \tag{8.14}$$

where $\mu_i^{\text{o,id(g)}}$ is the chemical potential of component i of an ideal gas phase in a standard state, i.e. standard chemical potential ($P = P^\circ = 1$ bar).

From the definition of μ_i Eq. (7.59),

$$\mu_1 = \left(\frac{\partial G}{\partial n_1}\right)_{T,P,n_2} \tag{8.15}$$

Therefore, by partially differentiating Eq. (8.14), the following relationship between chemical potential (μ_1) and partial pressure (Px_1) can be obtained:

$$\mu_1^{\text{id(g)}} = \left(\frac{\partial G}{\partial n_1}\right)_{T,P,n_2} = \mu_1^{\text{o,id(g)}} + RT\ \ln(Px_1) \tag{8.16}$$

where $\mu_1^{\text{id(g)}}$ is the chemical potential of Component 1 of an ideal gas phase.

In a similar manner, the following equation can be obtained:

$$\mu_2^{\text{id(g)}} = \left(\frac{\partial G}{\partial n_2}\right)_{T,P,n_1} = \mu_2^{\text{o,id(g)}} + RT\ \ln(Px_2) \tag{8.17}$$

According to Dalton's law, the partial pressure of Component 1 (P_1) is $P_1 = Px_1$ for an ideal gas. Therefore, Eq. (8.16) may be rewritten as

$$\mu_1^{\text{id(g)}} = \mu_1^{\text{o,id(g)}} + RT\ \ln P_1 \tag{8.18}$$

or

$$\mu_1^{\text{id(g)}} = \mu_1^{\text{o,id(g)}} + RT\ \ln P + RT\ \ln x_1 \tag{8.19}$$

As shown in Eq. (8.5), the term $\mu_1^{\text{o,id(g)}} + RT\ \ln P$ in Eq. (8.19) is equal to $\mu_1^{*,\text{id(g)}}$. Therefore, Eq. (8.19) may be rewritten as

$$\mu_1^{\text{id(g)}} = \mu_1^{*,\text{id(g)}} + RT\ \ln x_1 \tag{8.20}$$

As can be seen from Eq. (8.5), $\mu_1^{*,\text{id(g)}}$ is a function of temperature and pressure.

The chemical potential of ideal gas mixtures can be related to partial pressure, P_i, and concentration, x_i, in a simple and clear way, as is shown in Eqs. (8.18) and (8.20).

8.1.2 *Ideal Solutions*

8.1.2.1 The Relationship Between Chemical Potential and Concentration

The thermodynamic treatment of an ideal solution is, in essence, treated in the same manner as an ideal gas. In fact, experiments and molecular-theoretical discussions have shown that the chemical potential of an ideal solution, $\mu_i^{\text{id(sln)}}$, can be described in the same form as that of an ideal gas mixture:

$$\mu_i^{\text{id(sln)}} = \mu_i^{*,(\ell)} + RT\ \ln x_i \tag{8.21}$$

where $\mu_i^{*,(\ell)}$ is the chemical potential of pure component i ($x_i = 1$). As can be seen from the equation for an ideal gas mixture (Eq. (8.5)), $\mu_1^{*,\text{id(g)}}(T, P) = \mu_1^{\text{o,id(g)}}(P^{\circ}) + RT\ \ln P$. $\mu_i^{*,(\ell)}$ is also a function of temperature, T, and pressure, P. $\mu_i^{*,(\ell)}$ is the chemical potential of pure component i at temperature T and pressure P. On the other hand, $\mu_i^{\text{o},(\ell)}$ is the standard chemical potential, i.e. the chemical potential of pure component i at temperature T and pressure P° (= 1 bar). However, the influence of pressure on chemical potential is negligible in most cases. Hence, $\mu_i^{*,(\ell)} \approx \mu_i^{\text{o},(\ell)}$, and Eq. (8.21) may be written as

$$\mu_i^{\mathrm{id(sln)}} = \mu_i^{\circ,(\ell)} + RT\ \ln x_i \tag{8.22}$$

Equation (8.21) can be derived as follows:

The relationship between the chemical potential of component i of an ideal solution, $\mu_i^{\mathrm{id(sln)}}$, and concentration x_i of the solution (i.e. Eq. 8.21) can be derived from the vapour pressure of the gas phase, which is in equilibrium with the solution. An ideal solution can be defined as one that follows Raoult's law, and so the following equation is true when the vapour is in equilibrium with the ideal solution:

$$P_i = x_i P_i^* \tag{8.23}$$

where P_i is the vapour pressure of component i of the liquid solution of mole fraction x_i, and P_i^* is the vapour pressure of pure liquid i at temperature T.

The following relationship between the chemical potential of component i of an ideal solution $\left(\mu_i^{\mathrm{id(sln)}}\right)$ and the chemical potential of the gas phase $\left(\mu_i^{(\mathrm{g})}\right)$ is true when the two phases are in equilibrium:

$$\begin{aligned}\mu_i^{\mathrm{id(sln)}} &= \mu_i^{(\mathrm{g})} \\ &= \mu_i^{\circ,\mathrm{id(g)}} + RT\ \ln P_i\end{aligned} \tag{8.24}$$

where $\mu_i^{\circ,\mathrm{id(g)}}$ is the standard chemical potential, i.e. the chemical potential of component i when the partial pressure of component i (P_i) is 1 bar ($P_i = 1$ bar, i.e. pressure in a standard state).

As $P_i = x_i P_i^*$, Eq. (8.24) can be rewritten as

$$\mu_i^{\mathrm{id(sln)}} = \mu_i^{\circ,\mathrm{id(g)}} + RT\ \ln P_i^* + RT\ \ln x_i \tag{8.25}$$

On the other hand, when pure liquid component i and component i in the vapour and solution are in equilibrium,

$$\mu_i^{*,(\ell)} = \mu_i^{(\mathrm{g})} = \mu_i^{\circ,\mathrm{id(g)}} + RT\ \ln P_i^* \tag{8.26}$$

where $\mu_i^{*,(\ell)}$ is the chemical potential of pure component i in the liquid phase, which is in equilibrium with the vapour at $P_i^* = 1$ bar (which corresponds to $\mu_i^{\mathrm{id(sln)}}$ when $x_i = 1$ in Eq. (8.25)), and $\mu_i^{(\mathrm{g})}$ is the chemical potential of component i in the gas phase, which is in equilibrium with pure liquid component i at $P_i^* = 1$ bar (see Fig. 8.1).

Hence, by substituting Eq. (8.26) into Eq. (8.25), we obtain

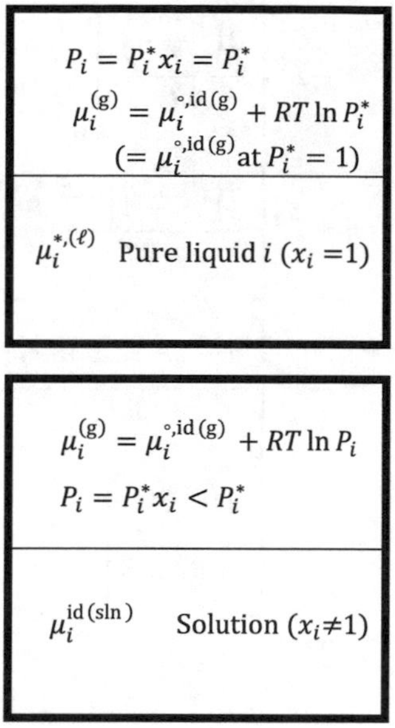

$P_i = P_i^* x_i = P_i^*$

$\mu_i^{(g)} = \mu_i^{\circ,\mathrm{id}\,(g)} + RT \ln P_i^*$

$(= \mu_i^{\circ,\mathrm{id}\,(g)}$ at $P_i^* = 1)$

$\mu_i^{*,(\ell)}$ Pure liquid i $(x_i = 1)$

$\mu_i^{(g)} = \mu_i^{\circ,\mathrm{id}\,(g)} + RT \ln P_i$

$P_i = P_i^* x_i < P_i^*$

$\mu_i^{\mathrm{id}\,(\mathrm{sln})}$ Solution $(x_i \neq 1)$

Fig. 8.1 Equilibrium between gas, pure liquid, and a solution

$$\mu_i^{\mathrm{id(sln)}} = \mu_i^{*,(\ell)} + RT \ln x_i \tag{8.27}$$

(End of the derivation of Eq. (8.21))

In the above discussion, gas phases are treated as ideal gases. In the case of a non-ideal gas, P_i and P_i^* are replaced by f_i and f_i^*, respectively, which are thermodynamic effective partial pressures, i.e. *fugacity*.

8.1.2.2 The Properties of an Ideal Solution

We can define a solution that satisfies Eq. (8.21) as an ideal solution. Based on this, several properties of an ideal solution can be derived.

8.1.2.3 Heat of Mixing

The heat of mixing, $\Delta_{\mathrm{mix}} H$, of an ideal solution is zero.

Let us substitute Eq. (8.21) into the *Gibbs-Helmholtz equation* (see Eq. (8.32)):

$$
\begin{aligned}
\overline{H}_i^{\mathrm{id(sln)}} &= -T^2\left[\frac{\partial\left(\frac{\mu_i^{\mathrm{id(sln)}}}{T}\right)}{\partial T}\right]_{P,x_i} \\
&= -T^2\left[\frac{\partial\left(\frac{\mu_i^{*,(\ell)}+RT\ \ln x_i}{\mathrm{T}}\right)}{\partial T}\right]_{P,x_i} \\
&= -T^2\left[\frac{\partial\left(\frac{\mu_i^{*,(\ell)}}{T}+R\ \ln x_i\right)}{\partial T}\right]_{P,x_i} \\
&= -T^2\left[\frac{\partial\left(\frac{\mu_i^{*,(\ell)}}{T}\right)}{\partial T}\right]_{P} \\
&= \overline{H}_i^{*,(\ell)}
\end{aligned}
\tag{8.28}
$$

where $\overline{H}_i^{\mathrm{id(sln)}}$ is the partial molar enthalpyof component$i$ at a given concentration, x_i; $\overline{H}_i^{*,(\ell)}$ is the partial molar enthalpy of pure component i; $\mu_i^{\mathrm{id(sln)}}$ is the chemical potential of component i of an ideal solution at a given concentration; $\mu_i^{*,(\ell)}$ is the chemical potential of pure component i in liquid form; T is the temperature; R is the gas constant; P is the pressure; and x_i is the mole fraction of component i.

Hence, the heat of mixing ($\Delta_{\mathrm{mix}}H$) for n_i moles is

$$
\Delta_{\mathrm{mix}}H = \sum_{i=1}^{r} n_i\left(\overline{H}_i^{\mathrm{id(sln)}} - \overline{H}_i^{*,(\ell)}\right) = 0 \tag{8.29}
$$

As $G = H - TS$, this relationship can also be derived using the following equation:

$$
\Delta_{\mathrm{mix}}H = \Delta_{\mathrm{mix}}G + T\Delta_{\mathrm{mix}}S \tag{8.30}
$$

By substituting Eqs. (8.12) and (8.13) into Eq. (8.30), we obtain

$$
\begin{aligned}
\Delta_{\mathrm{mix}}H &= RT(n_1 \ln x_1 + n_2 \ln x_2) - TR(n_1 \ln x_1 + n_2 \ln x_2) \\
&= 0
\end{aligned}
\tag{8.31}
$$

As can be seen based on Eq. (8.31), the heat of mixing, $\Delta_{\mathrm{mix}}H$, of an ideal solution is zero.

The *Gibbs-Helmholtz equation*,

$$
\left[\frac{\partial(G/T)}{\partial T}\right]_P = -\frac{H}{T^2} \tag{8.32}
$$

which is used to derive Eq. (8.31), can be derived as follows:

By differentiating $\frac{1}{T} \cdot G$ with respect to T at constant P,

$$\left[\frac{\partial\left(\frac{1}{T} \cdot G\right)}{\partial T}\right]_P = \frac{1}{T}\left(\frac{\partial G}{\partial T}\right)_P - \frac{1}{T^2}G \tag{8.33}$$

Using Eq. (7.31),

$$\left(\frac{\partial G}{\partial T}\right)_P = -S \tag{8.34}$$

where S is the entropy.

Therefore,

$$\left[\frac{\partial\left(\frac{1}{T} \cdot G\right)}{\partial T}\right]_P = -\frac{S}{T} - \frac{G}{T^2} = \frac{-ST - G}{T^2} \tag{8.35}$$

Hence, from the definition of G ($G \equiv H - TS$),

$$\left[\frac{\partial(G/T)}{\partial T}\right]_P = -\frac{H}{T^2} \tag{8.36}$$

(End of the derivation of the Gibbs-Helmholtz equation)

8.1.2.4 Volume Change of Mixing

The volume change of mixing, $\Delta_{\text{mix}} V$, of an ideal solution is zero.

Using Eq. (7.32), we obtain

$$\overline{V}_i^{\text{id(sln)}} = \left(\frac{\partial \mu_i^{\text{id(sln)}}}{\partial P}\right)_{T,x_i} \tag{8.37}$$

where $\overline{V}_i^{\text{id(sln)}}$ is the partial molar volume of component i at a given concentration, x_i; $\mu_i^{\text{id(sln)}}$ is the chemical potential of component i in an ideal solution at a given concentration, x_i; and P is the pressure.

By substituting Eq. (8.21) into the above equation,

$$\overline{V}_i^{(\text{sln})} = \left[\frac{\partial\left(\mu_i^{*,(\ell)} + RT\ \ln x_i\right)}{\partial P}\right]_{T,x_i}$$
$$= \left[\frac{\partial \mu_i^{*,(\ell)}}{\partial P}\right]_T$$
$$= \overline{V}_i^{*,(\ell)} \tag{8.38}$$

where $\overline{V}_i^{*,(\ell)}$ is the partial molar volume of pure component i and $\mu_i^{*,(\ell)}$ is the chemical potential of pure component i in liquid form.

Hence, the volume change of mixing, $\Delta_{\text{mix}}V$, for n_i moles is

$$\Delta_{\text{mix}}V = \sum_{i=1}^{r} n_i\left(\overline{V}_i^{\text{id(sln)}} - \overline{V}_i^{*,(\ell)}\right) = 0 \tag{8.39}$$

As can be seen in Eq. (8.39), the change in volume as a result of mixing, $\Delta_{\text{mix}}V$, of an ideal solution is zero.

8.2 Chemical Potential and Concentration of a Non-ideal Mixture

The relationship between the chemical potential and concentration/pressure of real solutions and gases cannot be described with simple forms such as Eqs. (8.18) and (8.21). Thus, we must consider the relationship between chemical potential and concentration for non-ideal gases and solutions.

8.2.1 Non-ideal Gases and Fugacity

8.2.1.1 Introduction of Fugacity

Up to approximately 1 bar, the chemical potential of real gases is approximated by Eqs. (8.18) and (8.19). However, strictly speaking, Eqs. (8.18) and (8.19) cannot be applied for real gases as they are.

Fugacity, f_i, is used to describe the chemical potential of non-ideal gases just as with that of ideal gases. Thus, the chemical potential of a non-ideal gas, $\mu_i^{\text{re(g)}}$, can be described as follows:

$$\mu_i^{\text{re(g)}} = \mu_i^{\text{o,re(g)}} + RT \ln f_i \tag{8.40}$$

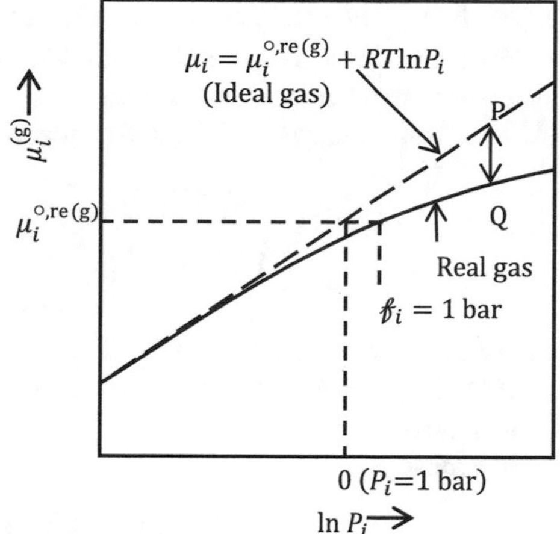

Fig. 8.2 Chemical potential of a gas in a standard state, $\mu_i^{\text{o,re(g)}}$

where $\mu_i^{\text{o,re(g)}}$ is the chemical potential when the fugacity of component i, f_i, is 1 bar, i.e. standard chemical potential.

Equation (8.40) must be true for the entire pressure range. A gas is ideal when the partial pressure of component i, P_i, $\to$ 0, meaning that Eq. (8.40) may be described as follows when $P_i \to 0$:

$$\mu_i^{\text{re(g)}} = \mu_i^{\text{o,re(g)}} + RT \ln P_i \tag{8.41}$$

and can also be described as

$$\mu_i^{\text{o,re(g)}} = \lim_{P_i \to 0} \left(\mu_i^{\text{re(g)}} - RT \ln P_i \right) \tag{8.42}$$

As shown in Fig. 8.2, $\mu_i^{\text{o,re(g)}}$ is chemical potential when it is assumed that a linear relationship between $\mu_i^{(g)}$ and ln P_i for ideal gases is the case up to $P_i = 1$ bar. As is discussed above, $\mu_i^{\text{o,re(g)}}$ (standard chemical potential) is equal to chemical potential at $f_i = 1$ for non-ideal gases.

8.2.1.2 Obtaining Fugacity

The fugacity of component i, f_i, can be obtained as follows:

According to Eq. (7.32), the relationship between the chemical potential of component i (pure gas), μ_i^*, and the molar volume of component i (pure gas), $\overline{V}_i^*$, is

$$\left(\frac{\partial \mu_i^*}{\partial P_i^*} \right)_T = \overline{V}_i^* \tag{8.43}$$

where P_i^* is the partial pressure of pure component i.

For a non-ideal gas, the ideal gas law, $\overline{V}_i^* = RT/P_i^*$, cannot be applied. The difference between an ideal gas and a non-ideal gas can be obtained by subtracting RT/P_i^* from both sides of the above equation:

$$\left(\frac{\partial \mu_i^{*,\mathrm{re(g)}}}{\partial P_i^*}\right)_T - \frac{RT}{P_i^*} = \overline{V}_i^{*,\mathrm{re(g)}} - \frac{RT}{P_i^*} \tag{8.44}$$

where $\mu_i^{*,\mathrm{re(g)}}$ is the chemical potential of component i (pure real gas), $\overline{V}_i^{*,\mathrm{re(g)}}$ is the molar volume of component i (pure real gas), R is the gas constant, and T is the temperature.

In addition,

$$\frac{RT}{P_i^*} = \left[\frac{\partial RT \ln P_i^*}{\partial P_i^*}\right]_T \tag{8.45}$$

Therefore, Eq. (8.44) can be described as

$$\left[\frac{\partial\left(\mu_i^{*,\mathrm{re(g)}} - RT \ln P_i^*\right)}{\partial P_i^*}\right]_T = \overline{V}_i^{*,\mathrm{re(g)}} - \frac{RT}{P_i^*} \tag{8.46}$$

By integrating Eq. (8.46) from 0 to P_i^* atm at constant T, we obtain

$$\left(\mu_i^{*,\mathrm{re(g)}} - RT \ln P_i^*\right) - \lim_{P_i^* \to 0}\left(\mu_i^{*,\mathrm{re(g)}} - RT \ln P_i^*\right) = \int_0^{P_i^*}\left(\overline{V}_i^{*,\mathrm{re(g)}} - \frac{RT}{P_i^*}\right)\mathrm{d}P_i^* \tag{8.47}$$

As shown in Eq. (8.42), the second term of the left side of the equation is

$$\lim_{P_i^* \to 0}\left(\mu_i^{*,\mathrm{re(g)}} - RT \ln P_i^*\right) = \mu_i^{\mathrm{o,re(g)}} \tag{8.48}$$

Hence, Eq. (8.47) becomes

$$\mu_i^{*,\mathrm{re(g)}} = \mu_i^{\mathrm{o,re(g)}} + RT \ln P_i^* + \int_0^{P_i^*}\left(\overline{V}_i^{*,\mathrm{re(g)}} - \frac{RT}{P_i^*}\right)\mathrm{d}P_i^* \tag{8.49}$$

Using Eq. (8.40), we obtain

$$f_i^* = \exp\left(\frac{\mu_i^{*,\mathrm{re(g)}} - \mu_i^{\mathrm{o,re(g)}}}{RT}\right) \tag{8.50}$$

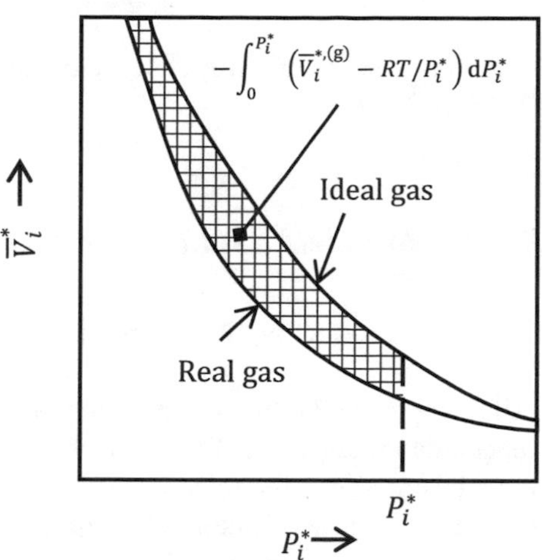

Fig. 8.3 Graphical integration performed to obtain the fugacity of a non-ideal gas (pure component i)

where f_i^* is the fugacity of component i (pure gas).

By substituting Eq. (8.49) into Eq. (8.50) we obtain

$$f_i^* = P_i^* \exp\left[\frac{1}{RT}\int_0^{P_i^*}\left(\overline{V}_i^{*,\mathrm{re(g)}} - \frac{RT}{P_i^*}\right)dP_i^*\right] \tag{8.51}$$

Hence, the chemical potential difference between ideal and non-ideal gases, PQ in Fig. 8.2, is equal to $-\int_0^{P_i^*}\left(\overline{V}_i^{*,\mathrm{re(g)}} - \frac{RT}{P_i^*}\right)dP_i^*$.

As is shown in Fig. 8.3, if the relationship between P_i^* and $\overline{V}_i^{*,(g)}$ of a non-ideal gas is known, the shaded area between the P_i^* versus $\overline{V}_i^{*,(g)}$ curve for ideal and non-ideal gases is equal to the integration term in Eq. (8.51) multiplied by −1. Therefore, if the state equation of a non-ideal gas is known, its fugacity, $\mu_i = \mu_i^{\circ} + RT\ln f_i$, can be obtained by substituting the value of the integration term, which can be obtained by graphical integration (as shown in Fig. 8.3).

Similarly, the chemical potential of a mixture of non-ideal gases of fixed composition can also be derived in the same way as for that of a pure gas, i.e.

$$\mu_i^{\mathrm{re(g)}} = \mu_i^{\circ,\mathrm{re(g)}} + RT\ln f_i \tag{8.52}$$

In this case, the fugacity of component i, f_i, in the above equation is

$$f_i = Px_i \exp\left[\frac{1}{RT}\int_0^{P}\left(\overline{V}_i^{\mathrm{re(g)}} - \frac{RT}{P}\right)dP\right] \tag{8.53}$$

where x_i is the mole fraction of component i.

$\mu_i^{\circ,\mathrm{re(g)}}$ in Eq. (8.52) is

$$\mu_i^{\text{o,re(g)}} = \lim_{P \to 0} \left[\mu_i^{*,\text{re(g)}} - RT \ln(Px_i) \right] \tag{8.54}$$

An in-depth description of fugacity can be found in Kirkwood and Oppenheim (1961, p. 92–94).

8.2.2 *Non-ideal Solutions and Activity*

8.2.2.1 Introduction of Activity

In the case of a real solution, it is rare to show the ideal solution behaviour in the whole concentration range, i.e. Eq. (8.21) is rarely satisfied. There are, however, several ways of describing the relationship between the chemical potential of component i, μ_i, and the concentration of component i, x_i, for a real solution. A convenient approach is to use the same form as for an ideal solution Eq. (8.21) and introduce the parameter of activity, a_i. Here, the chemical potential of a non-ideal solution, $\mu_i^{\text{re(sln)}}$, is described as

$$\mu_i^{\text{re(sln)}} = \mu_i^{\text{o,re(sln)}} + RT \ln a_i \tag{8.55}$$

$$a_i = f_i x_i \tag{8.56}$$

where x_i is the mole fraction of component i and f_i is the activity coefficient. As can be seen from this equation, $\mu_i^{\text{o,re(sln)}}$ is the chemical potential when $a_i = 1$, and is the chemical potential of component i in a standard state (i.e. standard chemical potential). The concept of activity was introduced by G. N. Lewis, and expanded the possible applications of the concept of chemical potential.

According to Eqs. (8.55) and (8.56),

$$a_i = \exp\left[\left(\mu_i^{\text{re(sln)}} - \mu_i^{\text{o,re(sln)}} \right) / RT \right] \tag{8.57}$$

$$f_i = \left\{ \exp\left[\left(\mu_i^{\text{re(sln)}} - \mu_i^{\text{o,re(sln)}} \right) / RT \right] \right\} / x_i \tag{8.58}$$

Therefore, activity, a_i, cannot be defined without defining the chemical potential in a standard state, $\mu_i^{\text{o,re(sln)}}$. In order to define the activity coefficient f_i, the unit of concentration (mole fraction or mass%, etc.) and standard chemical potential, $\mu_i^{\text{o,re(sln)}}$, must be defined.

Another way of describing the relationship between chemical potential and the concentration of a non-ideal solution is

$$\mu_i^{\text{re(sln)}} = \mu_i^{\text{o,re(sln)}} + \phi RT \ln x_i \tag{8.59}$$

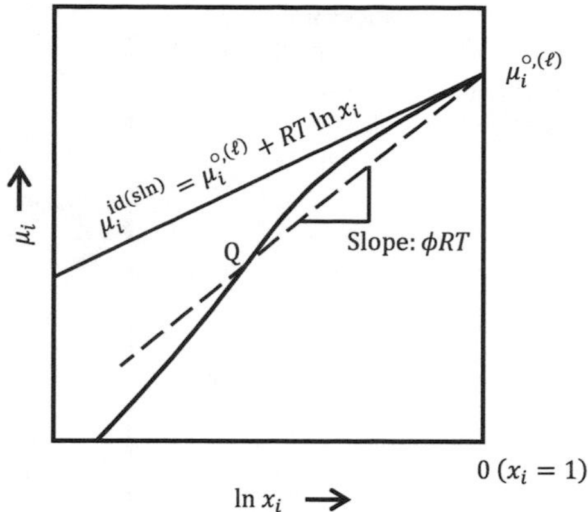

Fig. 8.4 Osmotic coefficient, ϕ

where ϕ is the osmotic coefficient. As is shown by the dashed line in Fig. 8.4, it is assumed that μ_i and $\ln x_i$ have a linear relationship, and the relationship between μ_i and $\ln x_i$ at Point Q is obtained by correcting the slope. It is convenient to use ϕ to discuss osmotic pressure, boiling-point elevation, and freezing-point depression.

8.2.2.2 Obtaining Activity

There are several methods of obtaining activity, using both experiments and calculations. The details of these methods are described in Sects. 8.3 and 8.4.

8.2.3 Choice of Standard Conditions and Unit of Concentration

8.2.3.1 Standard Conditions

A standard chemical potential is the value of chemical potential under specified standard conditions (i.e. the value in a standard state), and thus changes when conditions are altered. The standard conditions for activity and chemical potential are generally chosen in one of two ways.

Figure 8.5 shows the ideal solution behaviour of a non-ideal solution in two regions; a higher concentration region, and a dilute concentration region. The thinner of the two solid lines is the relationship between chemical potential (μ_i) and the natural logarithm of the mole fraction of component i($\ln x_i$) for an ideal solution, i.e. *Raoult's law*[2] is true for the entire concentration range.

---

[2]Raoult's law:

The partial pressure of a solute component i in a solution is proportional to its mole fraction.

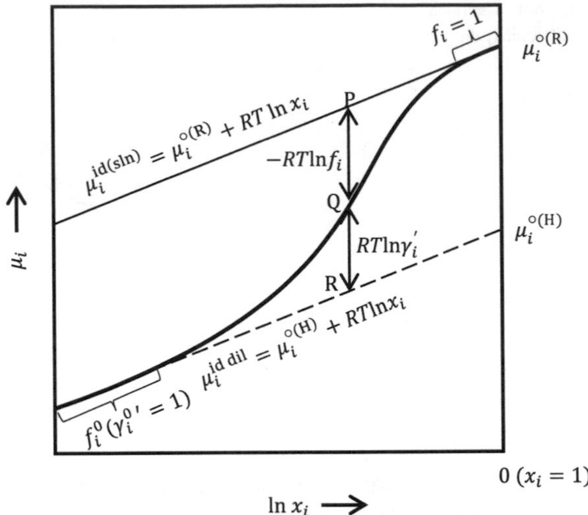

Fig. 8.5 The relationship between chemical potential, concentration, f_i, γ_i', $\mu_i^{\circ(R)}$, and $\mu_i^{\circ(H)}$

Therefore, one approach is to choose the pure liquid state of component i ($x_i = 1$) as a condition for the standard state, as its chemical potential, μ_i, can be described as follows using chemical potential under this standard condition, i.e. standard chemical potential, $\mu_i^{\circ(R)}$:

$$\begin{aligned}\mu_i &= \mu_i^{\circ(R)} + RT\ \ln a_i^{(R)} \\ &= \left(\mu_i^{\circ(R)} + RT\ \ln x_i\right) + RT\ \ln f_i \end{aligned} \tag{8.60}$$

where f_i is the activity coefficient (referenced to Raoult's law). $\mu_i^{\circ(R)}$ is the chemical potential of component i under the specified conditions (pure liquid) at the chosen standard state, and is called the *Raoultian standard state*. The superscripts ∘ and (R) denote 'standard state' and 'Raoultian standard state', respectively. $\mu_i^{\circ(R)}$ is the function of temperature and pressure, $a_i^{(R)}$ is the activity of the Raoultian standard state, R is the gas constant, T is the temperature, and x_i is the mole fraction of component i.

The term in brackets in Eq. (8.60) corresponds to the thinner line in Fig. 8.5, and $|RT\ \ln f_i|$ corresponds to $\overline{PQ}$ in Fig. 8.5.

Another approach to standard conditions is to choose the dilute concentration region which shows ideal solution behaviour, i.e. the region in which *Henry's law*[3] is true.

Assuming that the ideal solution behaviour of a non-ideal dilute solution is maintained up to $\ln x_i = 0$, i.e. $x_i = 1$ (pure i) (the dashed line in Fig. 8.5), the pure state of component i can be chosen as the standard state and the chemical potential

[3]Henry's law:

When the concentration of a solute component i is sufficiently low (i.e. it is sufficiently diluted), the partial pressure of solute component i in a solution is proportional to its mole fraction.

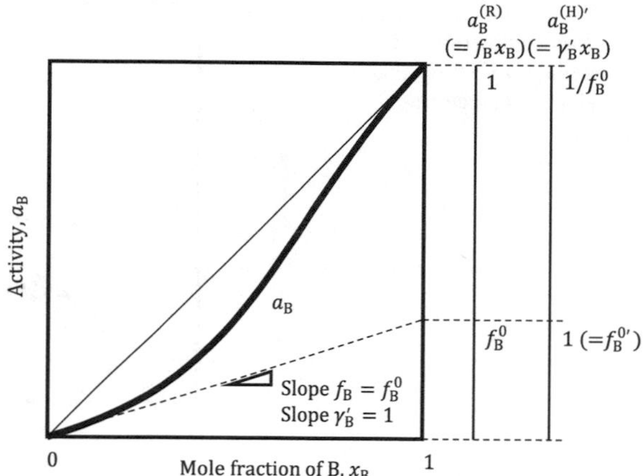

Fig. 8.6 Activity and composition (mole fraction), and the relationship between the Raoultian and Henrian standard states

at this point, $\mu_i^{\circ(H)}$, can be chosen as the standard chemical potential. In this case, chemical potential, μ_i, can be described as follows using chemical potential under this standard condition, i.e. standard chemical potential, $\mu_i^{\circ(H)}$:

$$\mu_i = \mu_i^{\circ(H)} + RT\ \ln\left(\gamma'_i x_i\right) \tag{8.61}$$

where γ'_i is the activity coefficient (referenced to Henry's law). $\mu_i^{\circ(H)}$ is the chemical potential of component *i* under the specified conditions in the standard state, called the *Henrian standard state* and denoted by the superscript (H). $\left|RT\ \ln \gamma'_i\right|$ in the above equation corresponds to $\overline{\text{QR}}$ in Fig. 8.5.

The activities of B of an A–B binary system are shown in Fig. 8.6 for a Raoultian standard state, $a_B^{(H)'}(= \gamma'_B x_B)$, and Henrian standard state, $a_B^{(H)'}(= \gamma'_B x_B)$.

As can be seen, the slope of the tangent line at $x_B \to 0$ (infinite dilute solution) of the activity curve, f_B^0, is the activity coefficient of this infinite dilute solution in a Raoultian standard state. In a Henrian standard state, the corresponding activity coefficient γ'_B is 1.

In the case of metallurgical refining processes, the system generally consists of the base metal as a solvent and a small amount of impurities and/or alloying constituents as solutes. In such a scenario, it is convenient to choose the Henrian standard state for the solutes due to the fact that only a small percentage (trace level) of oxygen, nitrogen, hydrogen, sulphur, etc. dissolves into metal. In addition, the pure liquid standard condition of these components does not exist in practice, as the liquid states of these components are not stable at around 1 bar and elevated temperatures. Therefore, there would be little necessity to adhere to μ_i°. $\mu_i^{\circ(H)}$ differs from $\mu_i^{\circ(R)}$, and $\mu_i^{\circ(H)}$ depends not only on temperature and pressure but also the properties of the solvent(s).

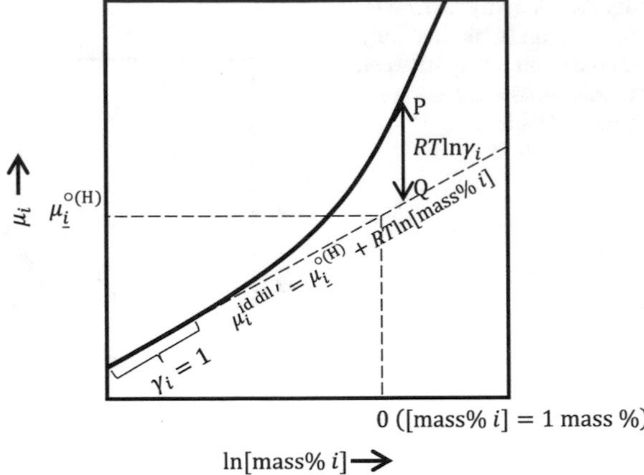

Fig. 8.7 Chemical potential and composition (natural logarithm of mass% i), and the relationship between $\mu_{\underline{i}}^{\circ(\mathrm{H})}$ and γ_i

8.2.3.2 Unit of Concentration

There are two primary ways of choosing unit of concentration.

One is mole fraction, x_i, which is explored in relation to high concentration regions and theoretical discussions. Mole fraction is generally used alongside standard chemical potential for Raoultian standard state, $\mu_i^{\circ(\mathrm{R})}$, wherein activity and activity coefficient are represented by $a_i^{(\mathrm{R})}$ and f_i, respectively.

Another common unit is mass percent of component i ([mass% i]), which is commonly used in practice in combination with standard chemical potential for *Henrian (1 mass% i) standard state*, $\mu_{\underline{i}}^{\circ(\mathrm{H})}$, which differs from the above-mentioned $\mu_i^{\circ(\mathrm{H})}$ (the underbar denotes 'dissolved'; $\underline{i}$ means 'dissolved component i'). In the case of a dilute solution, there is a region in which μ_i and ln[mass% i] have a linear relationship with slope RT, as shown in Fig. 8.7.

Assuming that this linear relationship is true up to [mass% i] = 1 mass% (i.e. ln[mass% i] = 0), the state of component i at [mass% i] = 1 mass% is chosen as a standard condition (Fig. 8.8).

The chemical potential at [mass% i] = 1 mass% can be set as a standard chemical potential, $\mu_{\underline{i}}^{\circ(\mathrm{H})}$. In this case, activity and the activity coefficient are represented by $a_i^{(\mathrm{H})}$ and γ_i, respectively. Chemical potential, μ_i, is expressed as

$$\mu_i = \mu_{\underline{i}}^{\circ(\mathrm{H})} + RT\ \ln a_i^{(\mathrm{H})} = \left[\mu_{\underline{i}}^{\circ(\mathrm{H})} + RT\ \ln[\mathrm{mass\%}\ i]\right] + RT\ \ln\gamma_i \tag{8.62}$$

The term in the square brackets in Eq. (8.62) corresponds to the dashed line in Fig. 8.7, and $|RT\ \ln\gamma_i|$ corresponds to $\overline{\mathrm{PQ}}$ in Fig. 8.7. The above-mentioned parameters are summarised in Table 8.1.

The values for activity and the activity coefficient are changed by the standard condition and choice of unit of concentration. The following equations relate f_i

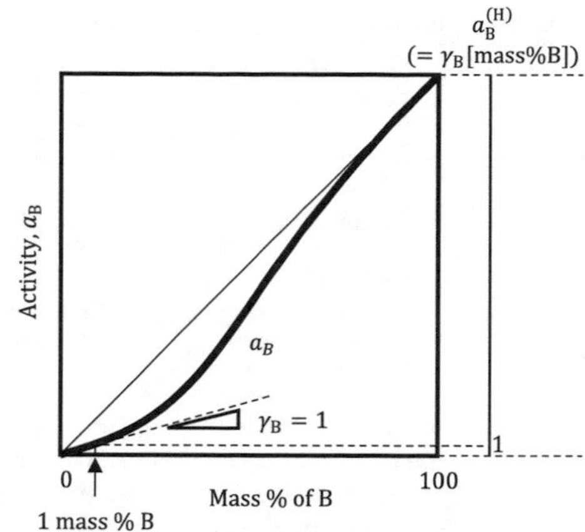

Fig. 8.8 Activity and composition (mass%). Henrian (1 mass% i) standard state

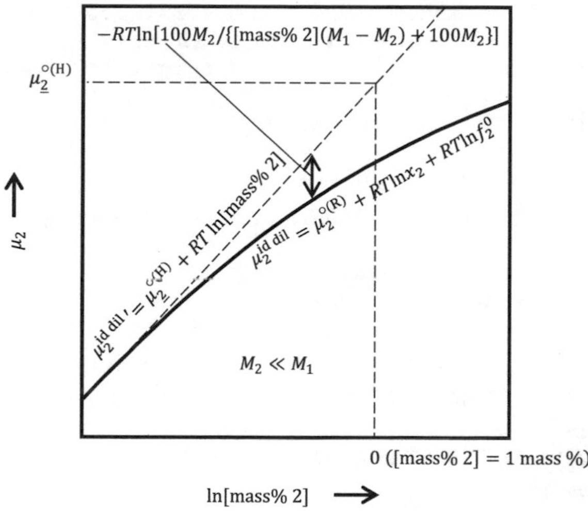

Fig. 8.9 The relationship between $\mu_2^{\text{id dil}}$ and $\mu_2^{\text{id dil}'}$

by Convention (a) and γ_i by Convention (b), and can be converted mutually (see Examples 8.1 and 8.2 for the derivation).

$$\gamma_i = \frac{\sum_{k=1}^{r} x_k M_k}{M_1} \cdot \frac{f_i}{f_i^0} \tag{8.71}$$

$$a_i^{(H)} = \frac{100M_i}{M_1 f_i^0} a_i^{(R)} \tag{8.72}$$

where

γ_i : Activity coefficient (Henrian (1 mass% i) standard state, concentration in mass%)

x_k : Mole fraction of component k

Table 8.1 Summary of parameters for different standard conditions

Convention	Standard condition	Standard chemical potential	Unit of concentration	Activity coefficient	Activity
(a)	Pure component i	$\mu_i^{\circ(R)}$	x_i Mole fraction	f_i	$a_i^{(R)}$
(b)	The state of component i of a solution at 1 mass% i[a]	$\mu_{\underline{i}}^{\circ(H)}$ (note that this is not $\mu_i^{\circ(H)}$)	[mass% i] Mass percent of component i	γ_i	$a_i^{(H)}$

[a]To be precise, this is the state of component i of a solution at [mass% i] = 1 mass% when it is 'assumed' that the linear relationship between μ_i and ln[mass% i] is true up to [mass% i] = 1 mass%. Depending on the solution, the linear relationship is not true even in an ideal dilute solution region when Henry's law is true, e.g. in a 1–2 component system

$x_2 = [\text{mass\% } 2] \cdot M_1/[[\text{mass\% } 2](M_1 - M_2) + 100M_2]$ (8.63)

Therefore, if the atomic weight or molecular weight of Components 1 and 2 (M_1 and M_2) differ markedly, $x_2 \propto$ [mass% 2] is not true, even if [mass% 2] < 1 mass%. As a result, even in an ideal dilute solution region, i.e. a concentration range that satisfies the equation for Raoultian standard state when $x_2 \to 0$

$\mu_2^{\text{id dil}} = \mu_2^{\circ(R)} + RT \ln x_2 + RT \ln f_2^0$ (8.64)

the following equation

$\mu_2^{\text{id dil}\prime} = \mu_{\underline{2}}^{\circ(H)} + RT \ln[\text{mass\% } 2]$ (8.65)

where f_2^0 is the activity coefficient of Component 2 when $x_2 \to 0$ (infinite dilute state), is no longer true.

Let us consider such a case.

Equation (8.64) is true for an ideal dilute solution.

By substituting Eq. (8.63) into Eq. (8.64), we obtain

$\mu_2 = \mu_2^{\circ(R)} + RT \ln f_2^0 + RT \ln\left[\frac{M_1}{[\text{mass\% } 2](M_1 - M_2) + 100M_2}\right] + RT \ln[\text{mass\% } 2]$ (8.66)

By substituting the following equation into Eq. (8.66) (see Example 8.1 for the derivation of Eq. (8.67))

$\mu_{\underline{2}}^{\circ(H)} = \mu_2^{\circ(R)} + RT \ln f_2^0 + RT \ln\left(\frac{M_1}{100M_2}\right)$ (8.67)

we obtain

$\mu_2 = \mu_{\underline{2}}^{\circ(H)} + RT \ln[\text{mass\% } 2] + RT \ln\left[\frac{100M_1}{[\text{mass\% } 2](M_1 - M_2) + 100M_2}\right]$ (8.68)

On the other hand

$\mu_2 = \mu_{\underline{2}}^{\circ(H)} + RT \ln[\text{mass\% } 2] + RT \ln \gamma_2$ (8.69)

By comparing Eqs. (8.68) and (8.69), we obtain

$\gamma_2 = \frac{100M_1}{[\text{mass\% } 2](M_1 - M_2) + 100M_2} \neq 1$ (8.70)

which implies that Eq. (8.65) is not true.

This relationship is shown in Fig. 8.9

f_i : Activity coefficient (Raoultian standard state, concentration in mole fraction)
f_i^0 : Activity coefficient when $x_i \to 0$ (Raoultian standard state, concentration in mole fraction)
M_i : Atomic/molecular weight of component i
$a_i^{(H)}$: Activity (Henrian (1 mass% i) standard state, concentration in mass%)
$a_i^{(R)}$: Activity (Raoultian standard state, concentration in mole fraction)

Example 8.1

Derive the relationships between standard chemical potential (Raoultian standard state), $\mu_i^{\circ(R)}$, and standard chemical potential (Henrian (1 mass% i) standard state) $\mu_{\underline{i}}^{\circ(H)}$; and between f_i and γ_i.

Solution:

If we consider Solvent 1 and the solution to have r components, the relationship between x_i and [mass% i] is

$$[\text{mass\% } i] = \frac{100 x_i M_i}{\sum_{k=1}^{r} x_k M_k} \tag{1}$$

where M_i is the atomic weight and x_i is the mole fraction of component i, respectively.

Chemical potential, μ_i, can be described as follows:

$$\mu_i = \mu_i^{\circ(R)} + RT\ \ln(f_i x_i) \tag{2}$$

$$= \mu_{\underline{i}}^{\circ(H)} + RT\ \ln(\gamma_i[\text{mass\% } i]) \tag{2'}$$

By substituting Eq. (1) into Eq. (2),

$$\mu_i = \mu_{\underline{i}}^{\circ(H)} + RT\ \ln\gamma_i + RT\ln\left(\frac{100 x_i M_i}{\sum_{k=1}^{r} x_k M_k}\right) \tag{3}$$

When [mass% 1] → 100 mass% (i.e. $[\text{mass\% } i]_{(i=2,3,\cdots,r)} \rightarrow 0$ mass%)

$$\gamma_i \rightarrow 1,\ (i = 2, 3, \cdots, r)$$

and when $x_1 \rightarrow 1$ (i.e. $x_{i\,(i\neq 1)} \rightarrow 0$)

$$f_i \rightarrow f_i^0,\ (i = 2, 3, \cdots, r)$$

For f_i^0, refer to Fig. 10.2.

According to Eqs. (2) and (3),

$$\mu_i^{\circ(R)} + RT\ \ln f_i^0 = \mu_{\underline{i}}^{\circ(H)} + RT\ln\left(\frac{100 M_i}{M_1}\right)$$

Therefore,

$$\mu_{\underline{i}}^{\circ(H)} = \mu_i^{\circ(R)} + RT\ \ln f_i^0 + RT\ln\left(\frac{M_1}{100 M_i}\right) \tag{4}$$

By substituting Eqs. (1), (2'), and (4) into Eq. (2), we obtain

$$\gamma_i = \frac{\sum_{k=1}^{r} x_k M_k}{M_i} \cdot \frac{f_i}{f_i^0} \tag{5}$$

Equations (4) and (5) are the relationships.

Example 8.2

Derive the relationship between $a_i^{(R)}$ and $a_i^{(H)}$.

Solution:

$$\begin{aligned}\mu_i &= \mu_i^{\circ(R)} + RT \ln a_i^{(R)} \\ &= \mu_{\underline{i}}^{\circ(H)} + RT \ln a_i^{(H)}\end{aligned} \tag{1}$$

From Example 8.1,

$$\mu_{\underline{i}}^{\circ(H)} = \mu_i^{\circ(R)} + RT \ln f_i^0 + RT \ln\left(\frac{M_1}{100M_i}\right)$$

By substituting this equation into Eq. (1), we obtain

$$\mu_i^{\circ(R)} + RT \ln a_i^{(R)} = \mu_i^{\circ(R)} + RT \ln f_i^0 + RT \ln\left(\frac{M_1}{100M_i}\right) + RT \ln a_i^{(H)}$$

Hence,

$$a_i^{(H)} = \frac{100M_i}{M_1 f_i^0} a_i^{(R)}$$

8.2.3.3 Other Standard Conditions

In addition to the above-mentioned standard conditions, a stable pure solid at a specified temperature can be chosen as a standard condition when a solution has a solubility limit and a solid phase crystallises in a pure form. Let us now consider the relationships between activity and activity coefficients when a pure solid is chosen as a standard condition and when a pure liquid is chosen as a standard condition.

Chemical potential, μ_i, can be written in terms of the chemical potential of pure component i (liquid), $\mu_i^{*,(\ell)}$, and the chemical potential of pure component i (solid), $\mu_i^{*,(s)}$, at temperature T in the form

Fig. 8.10 Pure solid and pure liquid standard states

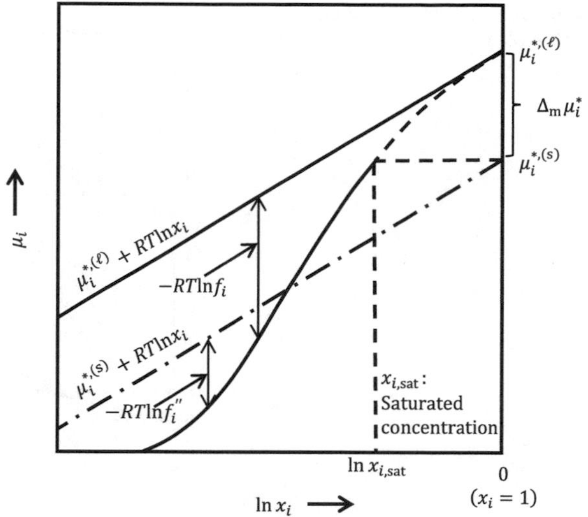

$$\begin{aligned} \mu_i &= \mu_i^{*,(\ell)} + RT \ln a_i \text{ (pure liquid is the standard tate)} \\ &= \mu_i^{*,(s)} - \mu_i^{*,(s)} + \mu_i^{*,(\ell)} + RT \ln a_i \\ &= \mu_i^{*,(s)} + RT \ln a_i'' \text{ (pure solid is the standard state)} \end{aligned} \tag{8.73}$$

The change in chemical potential due to the melting of pure component i, $\Delta_m \mu_i^*$, is defined as follows:

$$\Delta_m \mu_i^* \equiv \mu_i^{*,(\ell)} - \mu_i^{*,(s)} \tag{8.74}$$

We can then obtain the following equations:

$$\Delta_m \mu_i^* + RT \ln a_i = RT \ln a_i'' \tag{8.75}$$

$$\Delta_m \mu_i^* + RT \ln f_i = RT \ln f_i'' \tag{8.76}$$

where a_i'' and f_i'' are the activity and the activity coefficient, respectively, when a pure solid is chosen as a standard state.

The relationship between f_i'' and f_i is shown in Fig. 8.10. $\Delta_m \mu_i^* > 0$ at $T < T_m$, therefore it is always the case that $f_i'' > f_i$, according to Fig. 8.10. $\Delta_m \mu_i^*$ can be obtained using Eq. (8.109), the melting point of pure component $i(T_m)$, and the enthalpy of melting (enthalpy change due to melting), $\Delta_m \overline{H}_i^*$ at T_m. As a result, it is possible to perform a mutual conversion between a_i and a_i'', as well as between f_i and f_i''. This is shown in Eqs. (8.75) and (8.76), respectively.

An example of activity is shown in Fig. 8.11, wherein a pure solid is chosen as a standard state; the activity change of carbon in molten iron is a function of carbon concentration. Based on this, a_C'' negatively deviates from Raoult's law in the lower

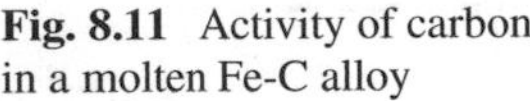

Fig. 8.11 Activity of carbon in a molten Fe-C alloy

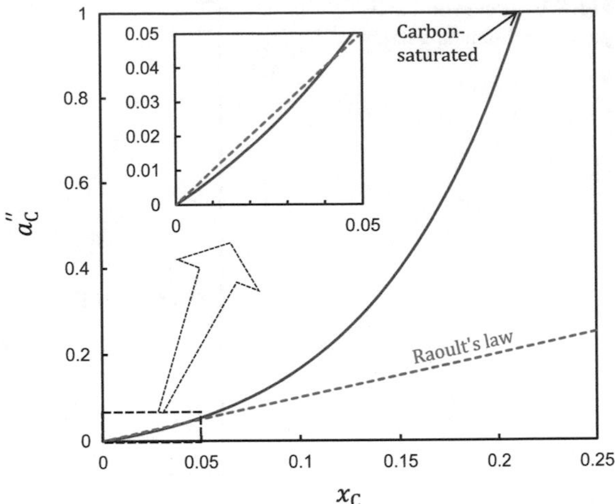

concentration region of carbon, and positively deviates in the higher concentration region. However, discussing such behaviour based on Raoult's law is, in essence, quite pointless, as the standard state of activity is a pure solid. The a''_{C} values of the Fe-C system relate to the fact that μ_i is below the line $\mu_{\mathrm{C}}^{*,(s)} + RT\ \ln x_i$ in the lower concentration region of carbon (therefore $f''_i < 1$), and above the line in the higher concentration region (therefore $f''_i > 1$), just as in the relationship between μ_i and $\ln x_i$ in Fig. 8.10.

Example 8.3

Take molality, m_i (number of moles of solute component i per 1 kg of solvent (Component 1), $\left[\mathrm{mol \cdot kg^{-1}}\right]$) as a unit of concentration and assume that the linear relationship $\mu_i = \mu_{\underline{i}}^{\circ(\mathrm{m})} + RT\ \ln m_i$, which is true for an infinite dilute state, is true up to $m_i = 1$; take the state of component i at $m_i = 1$ as a standard state. The activity and activity coefficient under these conditions are $a_i^{(\mathrm{m})}$ and $f_i^{(\mathrm{m})}$, respectively. Derive the relationship between $\mu_{\underline{i}}^{\circ(\mathrm{m})}$ and $\mu_i^{\circ(\mathrm{R})}$, as well as that between $f_i^{(\mathrm{m})}$ and f_i.

Solution:

The relationship between x_i and m_i is

$$x_i = \frac{M_1}{1000} \cdot \frac{m_i}{1 + (M_1/1000)\sum_{k=2}^{r} m_k}$$

$$\mu_i = \mu_i^{\circ(\mathrm{R})} + RT\ \ln f_i x_i \tag{1}$$

$$= \mu_i^{\circ(\mathrm{R})} + RT\ \ln f_i + RT\ \ln \frac{M_1 m_i}{1000\left\{1 + (M_1/1000)\sum_{k=2}^{r} m_k\right\}} \tag{2}$$

$$= \mu_{\underline{i}}^{\circ(\mathrm{m})} + RT\ \ln\left(f_i^{(\mathrm{m})} m_i\right) \tag{3}$$

When $x_i \to 0$(i.e. $x_1 \to 1$), $f_i = f_i^0$ and $f_i^{(\mathrm{m})} = 1$
Therefore,

$$\mu_i^{\circ(\mathrm{R})} + RT\ \ln f_i^0 + RT\ \ln\left(\frac{M_1}{1000}\right) = \mu_{\underline{i}}^{\circ(\mathrm{m})} \tag{4}$$

From Eqs. (1) and (4)

$$\mu_i = \mu_i^{\circ(\mathrm{R})} + RT\ \ln f_i + RT\ \ln\left(\frac{M_1}{1000}\right) + RT\ \ln \frac{m_i}{1 + (M_1/1000)\sum_{k=2}^{r} m_k} \tag{5}$$

From Eqs. (2) and (4),

$$\mu_i = \mu_i^{\circ(\mathrm{R})} + RT\ \ln f_i^0 + RT\ \ln\left(\frac{M_1}{1000}\right) + RT\ \ln f_i^{(\mathrm{m})} + RT\ \ln m_i \tag{6}$$

By comparing Eqs. (5) and (6), we obtain

$$RT\ \ln \frac{f_i}{1 + (M_1/1000)\sum_{k=2}^{r} m_k} = RT\ \ln\left(f_i^0 \cdot f_i^{(\mathrm{m})}\right)$$

Hence,

$$f_i^{(\mathrm{m})} = \frac{f_i}{f_i^0} \cdot \frac{1}{1 + (M_1/1000)\sum_{k=2}^{r} m_k} \tag{7}$$

Equations (4) and (7) are the relationships between $\mu_{\underline{i}}^{\circ(\mathrm{m})}$ and $\mu_i^{\circ(\mathrm{R})}$ and $f_i^{(\mathrm{m})}$ and f_i, respectively.

Example 8.4
When the common concentration unit x_i and chemical potential standards $\mu_i^{\circ(\mathrm{R})}$ and $\mu_i^{\circ(\mathrm{H})}$ are used, the activity coefficients are f_i and f_i', respectively. Derive the relationships between $\mu_i^{\circ(\mathrm{R})}$ and $\mu_i^{\circ(\mathrm{H})}$ and f_i and f_i' in manner similar to that used in Example 8.1.
Solution:

$$\begin{aligned}\mu_i &= \mu_i^{\circ(\mathrm{R})} + RT\ \ln(f_i x_i)\\ &= \mu_i^{\circ(\mathrm{H})} + RT\ \ln\left(f_i' x_i\right)\end{aligned} \quad (1)$$

When $x_i \to 0,\ f_i = f_i^0$ and $f_i' = 1$.
By substituting these values into Eq. (1),

$$\mu_i^{\circ(\mathrm{R})} + RT\ \ln\left(f_i^0 x_i\right) = \mu_i^{\circ(\mathrm{H})} + RT\ \ln x_i$$

Therefore,

$$\mu_i^{\circ(\mathrm{R})} = \mu_i^{\circ(\mathrm{H})} - RT\ \ln f_i^0 \quad (2)$$

By substituting Eq. (2) into Eq. (1), we obtain

$$\mu_i^{\circ(\mathrm{H})} + RT\ \ln\left(\frac{f_i}{f_i^0} x_i\right) = \mu_i^{\circ(\mathrm{H})} + RT\ \ln\left(f_i' x_i\right)$$

Hence,

$$f_i = f_i^0 f_i' \quad (3)$$

Equations (2) and (3) are the relationships between $\mu_i^{\circ(\mathrm{R})}$ and $\mu_i^{\circ(\mathrm{H})}$ and f_i and f_i', respectively.

Example 8.5

When the common standard chemical potential, $\mu_i^{\circ(\mathrm{R})}$, and concentration units x_i and [mass% i] are used, the activity coefficients are f_i and γ_i', respectively. Derive the relationships between f_i and γ_i'.

Solution:

$$\begin{aligned}\mu_i &= \mu_i^{\circ(\mathrm{R})} + RT\ \ln(f_i x_i)\\ &= \mu_i^{\circ(\mathrm{R})} + RT\ \ln\left(\gamma_i'\ [\text{mass\%}\ i]\right)\end{aligned}$$

Therefore,

$$f_i x_i = \gamma_i' [\text{mass\%}\ i] = \frac{100 x_i M_i}{\sum_{k=1}^{r} x_k M_k} \gamma_i'$$

$$f_i = \frac{100M_i}{\sum_{k=1}^{r} x_k M_k}\gamma_i'$$

Example 8.6

Derive the relationship between $\mu_i^{\circ(\mathrm{H})}$ and $\mu_{\underline{i}}^{\circ(\mathrm{H})}$.

Solution:

$$\begin{aligned}\mu_i &= \mu_i^{\circ(\mathrm{H})} + RT\ \ln\left(f_i' x_i\right)\\ &= \mu_{\underline{i}}^{\circ(\mathrm{H})} + RT\ \ln(\gamma_i[\mathrm{mass}\%\ i])\\ &= \mu_{\underline{i}}^{\circ(\mathrm{H})} + RT\ \ln\gamma_i + RT\ \ln\frac{100x_i M_i}{\sum_{k=1}^{r} x_k M_k}\end{aligned}$$

when $x_i \to 0$(i.e. $x_1 \to 1$), $f_i' = 1$, and $\gamma_i = 1$.

Hence,

$$\mu_i^{\circ(\mathrm{H})} = \mu_{\underline{i}}^{\circ(\mathrm{H})} + RT\ \ln\frac{100M_i}{M_1}$$

8.3 Determining Activity—Measurement Methods

The relationship between activity and concentration must be known in order to ascertain the equilibrium position of a reaction system involving solutions. In chemical engineering, solutions are used in, for example, adsorption towers to adsorb gases using solvents and distillation towers to distil chemicals. In metallurgical processes, molten alloys and molten slag are common solutions at elevated temperatures. It is important to understand the concept and be proficient in the use of activity. Activity and the other thermodynamic properties of solutions can be dealt with using equilibrium theory thanks to the advances made by our predecessors in the field, and today we have reached the stage where they have a great deal of practical use. In this section, the methods and principles used to determine activity are discussed.

Even today, it is difficult to calculate activity with much accuracy using ab initio calculations, and so it is usually determined by experiments for each system. Through this approach, activity data is obtained not only for the solution at room temperature but for alloy systems at elevated temperatures, and this data is of great practical use. Several major principles of activity-measurement methods that are applicable to elevated temperatures are described below.

8.3.1 Electromotive Force Measurement

By measuring the electromotive force,E, of a concentration cell (one side of which is a solution of pure component i), activity, a_i, can be calculated using the following equation:

$$a_i = \mathrm{e}^{-EZ\mathcal{F}/RT} \tag{8.77}$$

where $\mathcal{F}$ is the Faraday constant $\left(= 9.64853 \times 10^4\ \mathrm{C \cdot mol^{-1}}\right)$, Z is the charge number of ions entering the cell reaction, R is the gas constant, and T is the temperature. The activity of many alloys have been measured using this method and today, oxygen concentration cells that consist of solid electrolyte (such as $ZrO_2 \cdot CaO$) are widely used to measure the activities of oxygen or other alloying elements in molten metals.

The measurement principle is as follows:

Consider the following galvanic cell:

$$i(\text{pure liquid})|i\mathrm{X}(\text{electrolyte})|i - \mathrm{X}(\text{solution})$$

Assuming that component i in the electrolyte has the form of i^{z+} ions, the total cell reaction is

$$i(\text{pure liquid}) = \underline{i}\ (i - \mathrm{X}\ \text{solution}) \tag{8.78}$$

The change in Gibbs energy of the above reaction per mole of i^{z+} ions is

$$\Delta G = \mu_i^{(\mathrm{sln})} - \mu_i^{*,(\ell)} \tag{8.79}$$

Using Eq. (8.60), we obtain

$$\mu_i^{(\mathrm{sln})} = \mu_i^{\circ} + RT\ \ln a_i \tag{8.80}$$

As is discussed above (Sect. 8.1.2.1), $\mu_i^{*,(\ell)} \approx \mu_i^{\circ,(\ell)}$. Therefore,

$$\Delta G = \mu_i^{(\mathrm{sln})} - \mu_i^{*,(\ell)} = \left(\mu_i^{\circ} + RT\ \ln a_i\right) - \mu_i^{*,(\ell)} = RT\ \ln a_i \tag{8.81}$$

$-\Delta G$ corresponds to the maximum effective work, and is equal to the electrical work, $EZ\mathcal{F}$, done by the system to its surroundings when 1 mol of i^{z+} ions is carried with electromotive force E. Thus,

$$\Delta G = -EZ\mathcal{F} = RT \ln a_i \tag{8.82}$$

where E is the electromotive force, Z is the charge number of the ions, and $\mathcal{F}$ is the Faraday constant.

8.3.2 *Using an Oxygen Concentration Cell Consisting of Solid Electrolyte*

The partial oxygen pressure, P_{O_2}, of a gas phase can be measured using the following schematically represented *oxygen concentration cell* consisting of solid electrolyte:

$$\mathrm{Pt}|\mathrm{O_2}(P_{0,\mathrm{O_2}})|\mathrm{O^{2-}}|\mathrm{O_2}(P_{\mathrm{O_2}})|\mathrm{Pt}$$

By taking the gas phase, the partial oxygen pressure (P_{0,O_2}) of which is known, as the reference electrode, ΔG, of the cell reaction,

$$\mathrm{O_2}(P_{0,\mathrm{O_2}}) \rightarrow \mathrm{O_2}(P_{\mathrm{O_2}})$$

can be written as follows:

$$\Delta G = \mu_{O_2} - \mu_{0,O_2} = RT \ln\left(\frac{P_{O_2}}{P^\circ_{O_2}}\right) \tag{8.83}$$

where μ_{0,O_2} is the chemical potential of the reference electrode.

Due to the fact that the oxygen ion is O^{2-}, $Z = 4$ for 1 mol of O_2.

Hence,

$$\Delta G = -4\mathcal{F}E = RT \ln\left(\frac{P_{O_2}}{P_{0,O_2}}\right) \tag{8.84}$$

As can be seen from this equation, partial oxygen pressure, P_{O_2}, can be obtained by measuring electromotive force, E, at temperature T.

8.3.3 *Vapour-Pressure Measurement*

Activity must be determined using the fugacity of the vapour that is in equilibrium with the solid phase or the liquid phase. However, aside from in certain special cases, the vapour pressure of components that are in equilibrium with a solution can be approximated using vapour pressure rather than fugacity at below atmospheric pressure. Hence, activity can be obtained using

$$a_i^{(\mathrm{R})} = f_i x_i = \frac{P_i}{P_i^*} \tag{8.85}$$

where $a_i^{(\mathrm{R})}$ is the activity, f_i is the activity coefficient, x_i is the mole fraction of component i, P_i is the vapour pressure of component i of the solution at concentration x_i, and P_i^* is the vapour pressure of pure component i. Here, the standard state of activity, a_i, is a pure state, and the unit of concentration is mole fraction. The activities of Cu in a Fe-Cu system and Zn in a Cu-Zn system, for example, can be obtained using this method.

The measurement principle (derivation of Eq. (8.85)) is as follows:

When the gas phase of component i is in equilibrium with a liquid phase, for example, the chemical potential of component i in the solution, $\mu_i^{\mathrm{re(sln)}}$, is equal to the chemical potential of component i in the gas phase, $\mu_i^{\mathrm{re(g)}}$:

$$\mu_i^{\mathrm{re(sln)}} = \mu_i^{\mathrm{re(g)}} \tag{8.86}$$

Each chemical potential can be expressed as follows (these equations are identical to Eqs. (8.60) and (8.41)):

$$\mu_i^{\mathrm{re(sln)}} = \mu_i^{\circ(\mathrm{R})} + RT\, \ln a_i^{(\mathrm{R})} \tag{8.87}$$

$$\mu_i^{\mathrm{re(g)}} = \mu_i^{\circ,\mathrm{re(g)}} + RT\, \ln P_i \tag{8.88}$$

where $\mu_i^{\circ(\mathrm{R})}$ is the standard chemical potential (Raoultian standard state) and $\mu_i^{\circ,\mathrm{re(g)}}$ is the standard chemical potential of the gas phase.

Hence,

$$\ln a_i^{(\mathrm{R})} - \ln P_i = \frac{\left(\mu_i^{\circ,\mathrm{re(g)}} - \mu_i^{\circ(\mathrm{R})}\right)}{RT} \tag{8.89}$$

For a pure liquid, $P_i = P_i^*$ and $a_i^{(\mathrm{R})} = 1$. Thus, Eq. (8.89) is reduced to

$$-\ln P_i^* = \frac{\left(\mu_i^{\circ,\mathrm{re(g)}} - \mu_i^{\circ(\mathrm{R})}\right)}{RT} \tag{8.90}$$

Using Eqs. (8.89) and (8.90), we obtain

$$\ln a_i^{(\mathrm{R})} - \ln P_i = -\ln P_i^* \tag{8.91}$$

Hence,

$$a_i^{(\mathrm{R})} = f_i x_i = \frac{P_i}{P_i^*} \tag{8.92}$$

(End of the derivation of Eq. (8.85))

As can be seen from Eq. (8.92), it is possible to calculate activity, $a_i^{(\mathrm{R})}$, based on vapour pressures.

8.3.4 Measurement of Partition Constants

Consider Phases 1 and 2, which are immiscible solvents, and the solute component *i*, which is dissolved into the two phases. At equilibrium,

$$\left(K_{\mathrm{D}}^{\circ}\right)_i = \frac{a_{i,1}}{a_{i,2}} \tag{8.93}$$

where $a_{i,1}$ and $a_{i,2}$ are the activity of component *i* in Phases 1 and 2, respectively. The subscripts 1 and 2 denote the properties of Phases 1 and 2, respectively. $\left(K_{\mathrm{D}}^{\circ}\right)_i$ is the partition constant[4] and depends on the temperature, the pressure, and the solvents, but is not a function of concentration.

If the value of $\left(K_{\mathrm{D}}^{\circ}\right)_i$ and activity of the solute in one solvent are known, the activity of the solute in another solvent can be determined using Eq. (8.93). Both Pb and Ag are immiscible with Fe, and thus the activity of component *i* (e.g. Si) in a Fe-*i* system can be determined from these systems.

The measurement principle (derivation of Eq. (8.93)) is as follows:

The principle is similar to that of measuring vapour pressure described above (Sect. 8.3.3).

At equilibrium, the chemical potential of component *i* in Phase 1, $\mu_{i,1}$, is equal to that of Phase 2, $\mu_{i,2}$. Therefore,

$$\mu_{i,1} = \mu_{i,2} \tag{8.94}$$

[4]The term *partition constant* is today recommended by IUPAC, as the older term *partition coefficient*was equivocal and used as a synonym for partition constant, *partition ratio* (synonymous with *distribution constant*), and *distribution ratio* (once called *distributioncoefficient* or *extraction coefficient*, but these terms are not considered to be synonymous with distribution ratio); these terms should not be confused.

According to the IUPAC Gold Book:

Partition constant

"The ratio of activity of a given species A in the extract to its activity in the other phase with which it is in equilibrium".

Partition ratio (Distribution constant):

"The ratio of the concentration of a substance in a single definite form, A, in the extract to its concentration in the same form in the other phase at equilibrium".

Distribution ratio:

"The ratio of the total analytical concentration of a solute in the extract (regardless of its chemical form) to its total analytical concentration in the other phase".

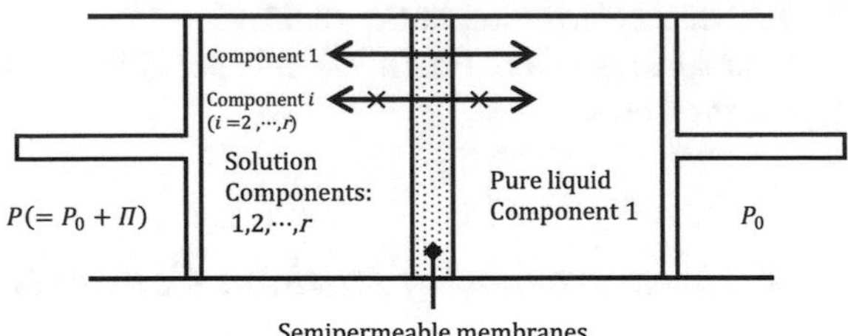

Fig. 8.12 A system consisting of a liquid solution that is separated from pure Liquid 1 (Component 1) by a membrane which is permeable to Component 1

$\mu_{i,1}$ and $\mu_{i,2}$, respectively, can be expressed as follows:

$$\mu_{i,1} = \mu_{\underline{i},1}^{\circ(\mathrm{H})} + RT\ \ln a_{i,1} \tag{8.95}$$

$$\mu_{i,2} = \mu_{\underline{i},2}^{\circ(\mathrm{H})} + RT\ \ln a_{i,2} \tag{8.96}$$

where $\mu_{\underline{i}}^{\circ(\mathrm{H})}$ is the standard chemical potential (Henrian (1 mass% i) standard state).

Thus,

$$\ln\left(\frac{a_{i,1}}{a_{i,2}}\right) = \frac{-\left(\mu_{\underline{i}}^{\circ(\mathrm{H})(1)} - \mu_{\underline{i}}^{\circ(\mathrm{H})(2)}\right)}{RT} \tag{8.97}$$

As shown in Eq. (10.6), in an equilibrium state the difference in the standard chemical potential is equal to ln K (K is the equilibrium constant, and here corresponds to the partition constant, $\left(K_{\mathrm{D}}^{\circ}\right)_i$).

Therefore,

$$\ln\left(\frac{a_{i,1}}{a_{i,2}}\right) = \frac{-\left(\mu_{\underline{i}}^{\circ(\mathrm{H})(1)} - \mu_{\underline{i}}^{\circ(\mathrm{H})(2)}\right)}{RT} = \ln\left(K_{\mathrm{D}}^{\circ}\right)_i \tag{8.98}$$

Hence,

$$\frac{a_i^{(1)}}{a_i^{(2)}} = \left(K_{\mathrm{D}}^{\circ}\right)_i \tag{8.99}$$

8.3.5 *Osmotic Pressure Measurement*

Consider a system that consists of a liquid solution of r components (Component 1, 2, …, r), separated from pure Liquid 1 (Component 1) by a non-deformable, heat-conducting membrane that is only permeable to Component 1 (Fig. 8.12). When the solution and pure liquid are in equilibrium, the activity of Component 1, a_1, can be obtained using the following equation:

$$RT \ln a_1 = -\Pi \overline{V}_1(P_0) \tag{8.100}$$

by measuring the osmotic pressure, $\Pi = P - P_0$.

P and P_0 are the pressure of the solution phase and the pure liquid phase, respectively; R is the gas constant, T is the temperature, and $\overline{V}_1(P_0)$ is the partial molar volume of Component 1 in the solution at pressure P_0.

The measurement principle is as follows:

The relationship between chemical potential μ_1 and partial molar volume $\overline{V}_1$ is described as follows (according to Eq. (7.32)):

$$\left(\frac{\partial \mu_1}{\partial P}\right)_{T,x} = \overline{V}_1 \tag{8.101}$$

Therefore, by integrating the equation from P_0 to $P_0 + \Pi$, we obtain

$$\mu_1(T, P, x) - \mu_1(T, P_0, x) = \int_{P_0}^{P_0+\Pi} \overline{V}_1(P)\,\mathrm{d}P \tag{8.102}$$

where x is the set of mole fractions, $x_1, x_2, \cdots, x_{r-1}$..

When the solution and pure liquid are in equilibrium, the chemical potential of Component 1 in the solution is equal to that of pure liquid:

$$\mu_1(T, P, x) = \mu_1^*(T, P_0) \tag{8.103}$$

Therefore, the left-hand side of Eq. (8.102) is

$$\mu_1(T, P, x) - \mu_1(T, P_0, x) = \mu_1^*(T, P_0) - \mu_1(T, P_0, x) \tag{8.104}$$

As $\mu_1 = \mu_1^* + RT \ln a_1$,

$$\mu_1^*(T, P_0) - \mu_1(T, P_0, x) = -RT \ln a_1 \tag{8.105}$$

Hence, the left-hand side of the equation, $\mu_1(T, P, x) - \mu_1(T, P_0, x)$, is equal to $-RT \ln a_1$.

By applying an incompressive approximation, we obtain

$$\overline{V}_1(P) = \overline{V}_1(P_0) \tag{8.106}$$

Thus, the right-hand side of Eq. (8.102) is

$$\int_{P_0}^{P_0+\Pi} \overline{V}_1(P)\mathrm{d}P = \int_{P_0}^{P_0+\Pi} \overline{V}_1(P_0)\mathrm{d}P = \left[P\overline{V}_1(P_0)\right]_{P_0}^{P_0+\Pi} = \Pi \overline{V}_1(P_0) \tag{8.107}$$

Hence,

$$-RT\ \ln a_1 = \Pi \overline{V}_1(P_0) \tag{8.108}$$

8.3.6 *Other Methods*

In addition to the above-mentioned methods, activity can be obtained through light-scattering measurements. This method can be applied to species of high molecular weight. In addition, the activity coefficient can be obtained using the equilibrium constant of a chemical reaction that is widely used to obtain the activity of low-concentration non-metallic elements such as $\underline{\mathrm{O}}$, $\underline{\mathrm{S}}$, $\underline{\mathrm{C}}$, etc. in molten alloys.

8.4 Determining Activity—Calculation Methods

8.4.1 *Calculation of Activity Using Equilibrium Phase Diagrams*

The activity of a solution component (composition at liquidus temperature T) can be obtained when the solid phase, which coexists with the liquid phase, is regarded as a pure phase.

8.4.1.1 Activity of a Solvent Component

Component 1 can be regarded as a solvent, as its solubility is high at temperature T, which is close to the melting point of pure Component 1 (T_m), as shown in Fig. 8.13.

Convention (a) can be used to describe activity. The activity of solvent Component 1 at temperature T and concentration x_1 can be obtained using the following equation, provided we have read the liquidus temperature T of Component 1 at pressure P from the phase diagram:

$$\ln a_1^{(\mathrm{R})} = -\left(\frac{\Delta_\mathrm{m}\overline{H}_1^*}{R}\right)\left(\frac{1}{T} - \frac{1}{T_\mathrm{m}}\right) \tag{8.109}$$

where $a_1^{(\mathrm{R})}$ is the activity of Component 1 (Raoultian standard state), $\Delta_\mathrm{m}\overline{H}_1^*$ is the molar heat of melting of pure Component 1 at the melting point T_m, R is the gas constant, and T is the temperature.

The principle is as follows:

On the liquidus line,

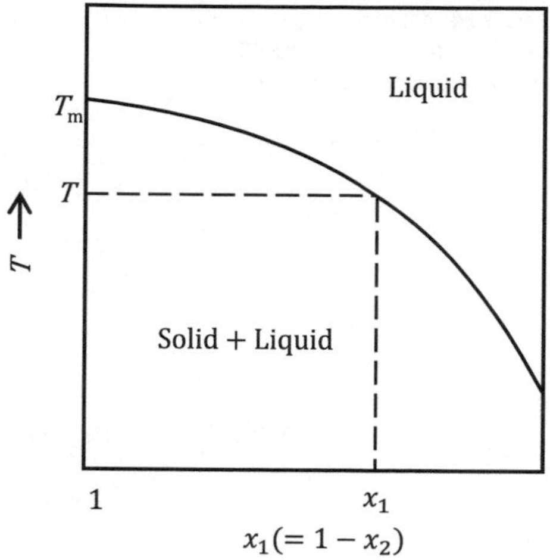

Fig. 8.13 Liquidus curve of a 1–2 binary system

$$\mu_1^{*,(s)} = \mu_1^{re(sln)} \tag{8.110}$$

where $\mu_1^{*,(s)}$ is the chemical potential of pure Component 1 in the solid phase and $\mu_1^{re(sln)}$ is the chemical potential of Component 1 in the solution (liquid phase).

This is true when the solid phase, which coexists with the liquid phase, can be regarded as a pure phase.

Using Eq. (8.60), we obtain

$$\mu_i^{re(sln)} = \mu_i^{o(R)} + RT \ln a_i^{(R)} \tag{8.111}$$

where $\mu_i^{o(R)}$ is the standard chemical potential (Raoultian standard state).

As can be seen from Fig. 8.14 and the definition $\Delta G \equiv \Delta H - T\Delta S$, a change in the chemical potential of pure Component 1 due to melting, $\Delta_m \mu_1^*$, can be described as

$$\Delta_m \mu_1^* = \mu_1^{*,(\ell)} - \mu_1^{*,(s)} = \Delta_m \overline{H}_1^* - T\Delta_m \overline{S}_1^* \tag{8.112}$$

At $T = T_m$,

$$\Delta_m \overline{H}_1^* - T_m \Delta_m \overline{S}_1^* = 0 \tag{8.113}$$

Hence,

$$\Delta_m \overline{S}_1^* = \frac{\Delta_m \overline{H}_1^*}{T_m} \tag{8.114}$$

Generally, the temperature dependence of $\Delta_m \overline{H}_1^*$ and $\Delta_m \overline{S}_1^*$ are small, and thus

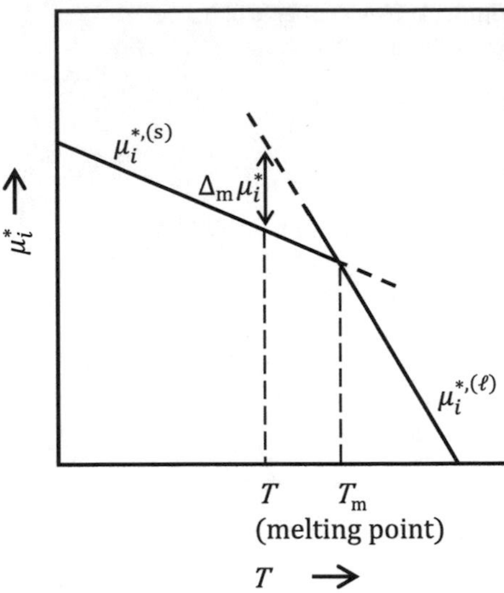

Fig. 8.14 Change in chemical potential as a result of melting at temperature T

$$\Delta_m \mu_1^* \approx \Delta_m \overline{H}_1^* - T \frac{\Delta \overline{H}_1^*}{T_m}$$
$$= \Delta_m \overline{H}_1^* \left(1 - \frac{T}{T_m}\right) \tag{8.115}$$

From Eqs. (8.110) and (8.111),

$$-RT \ln a_1^{(R)} = \mu_i^{\circ(R)} - \mu_1^{*,(s)} \tag{8.116}$$

Here, the chemical potential of pure Component 1 is approximately equal to that of its standard state, $\mu_i^{\circ(R)} \approx \mu_1^{*,(\ell)}$, as mentioned in Sect. 8.1.2.

Therefore,

$$-RT \ln a_1^{(R)} = \mu_i^{\circ(R)} - \mu_1^{*,(s)} = \mu_1^{*,(\ell)} - \mu_1^{*,(s)} = \Delta_m \overline{H}_1^* \left(1 - \frac{T}{T_m}\right) \tag{8.117}$$

Hence,

$$\ln a_1^{(R)} = -\left(\frac{\Delta_m \overline{H}_1^*}{R}\right)\left(\frac{1}{T} - \frac{1}{T_m}\right) \tag{8.118}$$

This method allows only activity (or the activity coefficient) to be obtained at a specific concentration and temperature. However, the temperature dependences of activity and the activity coefficient are small. Therefore, activity at a certain temperature is approximately equal to the activity obtained for another temperature. If the temperature coefficient must be taken into account, the following equation should be used:

$$\ln f_1 = \frac{\Delta_{\mathrm{mix}}\overline{H}_1}{RT} + \mathrm{C} \tag{8.119}$$

where $\Delta_{\mathrm{mix}}\overline{H}_1$ is the differential heat of solution of Component 1 (for a more in-depth discussion, see Chap. 9).

By ascertaining the activity coefficient, f_1, and $\Delta_{\mathrm{mix}}\overline{H}_1$ at temperature T on the liquidus line, the constant C can be defined. Hence, f_1 at the desired temperature can be calculated.

Example 8.7

Calculate the activity of silicon, a_{Si}, on the high-Si concentration side ($0.8 \le x_{\mathrm{Si}} \le 1$) of a Fe–Si alloy using the phase diagram (Fig. 8.15). $\Delta_{\mathrm{m}}\overline{H}^*_{\mathrm{Si}}$ is 50550 $\mathrm{J \cdot mol^{-1}}$.

Solution:

The melting point, T_{m}, of Si is 1703 K.

Therefore,

$$\begin{aligned}\ln a^{(\mathrm{R})}_{\mathrm{Si}} &= -\left(\frac{\Delta_{\mathrm{m}}\overline{H}^*_{\mathrm{Si}}}{R}\right)\left(\frac{1}{T} - \frac{1}{T_{\mathrm{m}}}\right)\\ &= -\frac{50550}{8.314}\left(\frac{1}{T} - \frac{1}{1703}\right)\\ &= -\frac{6080}{T} + 3.570\end{aligned}$$

Using this equation, the activity of Si $a^{(\mathrm{R})}_{\mathrm{Si}}$ at temperature T can be calculated. By reading concentration x_{Si} on the liquidus line, the activity coefficient, f_{Si}, can also be calculated using $a^{(\mathrm{R})}_{\mathrm{Si}} = f_{\mathrm{Si}}x_{\mathrm{Si}}$.

8.4.1.2 Activity of a Solute Component

When the solubility of Component 1 is extremely low ($x_1 \ll 1$) it can be regarded as a solute, and Convention (b) (Sect. 8.2.3) is used to describe its activity. By ascertaining the solubility of solute Component 1 at temperatures T([mass% 1 at T]) and T_0([mass% 1 at T_0]) based on the liquidus line, the activity or activity coefficient of Component 1 can be ascertained using the following equation and the value of molar heat of the dissolution of pure solid Component 1 into an infinite dilute solution $\left(\Delta\overline{H}^{\circ(\mathrm{H})}_1 = \overline{H}^{\circ(\mathrm{H})}_1 - \overline{H}^{*,(\mathrm{s})}_1\right)$.

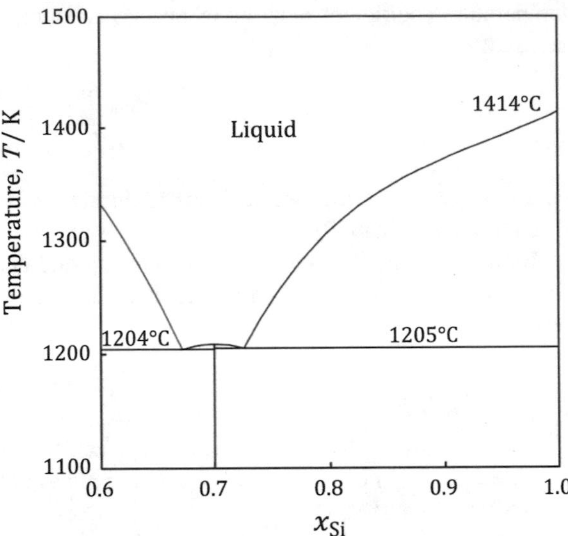

Fig. 8.15 Fe-Si phase diagram (calculated using Thermo-Calc ver. 3.1, TCBIN database)

$$\ln a_1 - \ln[\text{mass\% 1 at } T_0] = \int_{T_0}^{T} \frac{\Delta \overline{H}_1^{\circ(\mathrm{H})}}{RT^2} \mathrm{d}T \tag{8.120}$$

where $a_1 = \gamma_1[\text{mass\% 1}]$, $\overline{H}_1^{\circ(\mathrm{H})}$ is the partial molar enthalpy of Component 1 in the infinite dilute solution, and [mass% 1 at T_0] is very small. Thus, $\gamma_1 \approx 1$, $\overline{H}_1^{*,(\mathrm{s})}$ is the enthalpy of the pure Component 1 in the solid phase.

The principle (derivation of Eq. (8.120)) is as follows:

At temperature T, the chemical potential of pure solid component i is

$$\mu_1^{*,(\mathrm{s})}(T) = \mu_{\underline{1}}^{\circ(\mathrm{H})}(T) + RT\ \ln(\gamma_1[\text{mass\% 1 at } T]) \tag{8.121}$$

where $\mu_{\underline{1}}^{\circ(\mathrm{H})}$ is the standard chemical potential of Component 1 (Henrian (1 mass% i) standard state).

At temperature T_0,

$$\mu_1^{*,(\mathrm{s})}(T_0) = \mu_{\underline{1}}^{\circ(\mathrm{H})}(T_0) + RT_0 \ln[\text{mass\% 1 at } T_0] \tag{8.122}$$

Hence,

$$\ln(\gamma_1[\text{mass\% 1 at } T]) - \ln[\text{mass\% 1 at } T_0] = \frac{1}{R}\left(\frac{\mu_1^{*,(\mathrm{s})}(T) - \mu_{\underline{1}}^{\circ(\mathrm{H})}(T)}{T} - \frac{\mu_1^{*,(\mathrm{s})}(T_0) - \mu_{\underline{1}}^{\circ(\mathrm{H})}(T_0)}{T_0}\right) \tag{8.123}$$

According to the,

$$\frac{\overline{H}_i}{T^2} = -\left(\frac{\partial(\mu_i/T)}{\partial T}\right)_{P,x_i} \tag{8.124}$$

where $\overline{H}_i$ is the partial molar enthalpy of component i, T is the temperature, μ_i is the chemical potential of component i, P is the pressure, and x_i is the mole fraction of component i.

Therefore,

$$\begin{aligned}\int_{T_0}^{T} \frac{\mu_1^{*,(s)}(T)}{T^2}\mathrm{d}T &= -\int_{T_0}^{T} \mathrm{d}\left(\frac{\mu_1^{*,(s)}(T)}{T}\right) \\ &= -\left[\frac{\mu_1^{*,(s)}(T)}{T}\right]_{T_0}^{T} \\ &= -\frac{\mu_1^{*,(s)}(T)}{T} + \frac{\mu_1^{*,(s)}(T_0)}{T_0}\end{aligned} \tag{8.125}$$

Similarly,

$$\int_{T_0}^{T} \frac{\mu_{\underline{1}}^{\circ(\mathrm{H})}(T)}{T^2}\mathrm{d}T = -\frac{\mu_{\underline{1}}^{\circ(\mathrm{H})}(T)}{T} + \frac{\mu_{\underline{1}}^{\circ(\mathrm{H})}(T_0)}{T_0} \tag{8.126}$$

Equation (8.126) minus Eq. (8.125) is equal to the term $\frac{\mu_1^{*,(s)}(T)-\mu_{\underline{1}}^{\circ(\mathrm{H})}(T)}{T} - \frac{\mu_1^{*,(s)}(T_0)-\mu_{\underline{1}}^{\circ(\mathrm{H})}(T_0)}{T_0}$ in Eq. (8.123).

Hence,

$$\ln(\gamma_1[\text{mass\% 1 at } T]) - \ln[\text{mass\% 1 at } T_0] = \int_{T_0}^{T} \frac{\left(\overline{H}_1^{\circ(\mathrm{H})} - \overline{H}_1^{*,(s)}\right)}{RT^2}\mathrm{d}T \tag{8.127}$$

According to the Gibbs-Helmholtz equation,

$$\frac{\overline{H}_1^{\circ(\mathrm{H})} - \overline{H}_{\underline{1}}^{\circ(\mathrm{H})}}{T^2} = \left\{\frac{\partial\left[\left(\mu_{\underline{1}}^{\circ(\mathrm{H})} - \mu_1^{\circ(\mathrm{H})}\right)/T\right]}{\partial T}\right\}_{P,x_i} \tag{8.128}$$

and the relationship between $\mu_{\underline{1}}^{\circ(\mathrm{H})}$ and $\mu_1^{\circ(\mathrm{H})}$ is

$$\mu_{\underline{1}}^{\circ(\mathrm{H})} = \mu_1^{\circ(\mathrm{H})} + RT\,\ln\left(\frac{M_1}{100M_2}\right) \tag{8.129}$$

(cf. Example 8.6).

Therefore, $\left\{\frac{\partial\left[\left(\mu_{\underline{1}}^{\circ(\mathrm{H})} - \mu_1^{\circ(\mathrm{H})}\right)/T\right]}{\partial T}\right\}_{P,x_i}$ in Eq. (8.128) = 0, as

$$\left\{\frac{\partial\left[\left(\mu_{\underline{1}}^{\circ(\mathrm{H})}-\mu_{1}^{\circ(\mathrm{H})}\right)/T\right]}{\partial T}\right\}_{P,x_i}=\left\{\left(\frac{\partial\left[R\ln\left(\frac{M_1}{100M_2}\right)\right]}{\partial T}\right)\right\}_{P,x_i}$$
$$=0 \tag{8.130}$$

Thus,

$$\frac{\overline{H}_{1}^{\circ(\mathrm{H})}-\overline{H}_{\underline{1}}^{\circ(\mathrm{H})}}{T^2}=0 \tag{8.131}$$

Hence,

$$\overline{H}_{\underline{1}}^{\circ(\mathrm{H})}=\overline{H}_{1}^{\circ(\mathrm{H})} \tag{8.132}$$

Thus,

$$\Delta\overline{H}_{1}^{\circ(\mathrm{H})}=\overline{H}_{\underline{1}}^{\circ(\mathrm{H})}-\overline{H}_{1}^{*,(\mathrm{s})}=\overline{H}_{1}^{\circ(\mathrm{H})}-\overline{H}_{1}^{*,(\mathrm{s})} \tag{8.133}$$

Hence, Eq. (8.127) is identical to Eq. (8.120).
(End of the derivation of Eq. (8.120))

8.4.2 Calculating Activity Using the Gibbs-Duhem Equation

In the case of a homogeneous system, the Gibbs-Duhem equation (cf. Sect. 7.2.2) is true:

$$\sum_{i=1}^{r}x_i\mathrm{d}\mu_i+\overline{S}\mathrm{d}T-\overline{V}\mathrm{d}P=0 \tag{8.134}$$

where x_i is the mole fraction of component i, μ_i is the chemical potential of component i, $\overline{S}$ is the partial molar entropy, T is the temperature, $\overline{V}$ is the partial molar volume, and P is the pressure.

The Gibbs-Duhem equation is important in the thermodynamic treatment of solutions because it can be used to calculate the activity of binary and ternary systems. In the case of binary systems in particular, the calculation is relatively simple and has a high accuracy. Therefore, the Gibbs-Duhem equation is often used to calculate the activity of a component when the activities of other components are known.

In the case of a 1–2 binary system, Eq. (7.67) may be changed as follows at constant T and P:

$$x_1\mathrm{d}\mu_1+x_2\mathrm{d}\mu_2=0 \tag{8.135}$$

Thus,

$$d\mu_1 = \frac{x_2}{x_1} d\mu_2 \tag{8.136}$$

$\mu_i = \mu_i^\circ + RT \ln a_i$. Therefore, $d\mu_i = RT d \ln a_i$ ($i = 1,2$) at constant T and P. Hence,

$$d \ln a_1 = -\left(\frac{x_2}{x_1}\right) d \ln a_2 \tag{8.137}$$

However,

$$dx_1 = -dx_2 (\text{because } x_1 + x_2 = 1) \tag{8.138}$$

Thus,

$$x_1\left(\frac{dx_1}{x_1}\right) = -x_2\left(\frac{dx_2}{x_2}\right) \tag{8.139}$$

Therefore,

$$x_1 d \ln x_1 = -x_2 d \ln x_2 \left(\text{because } \int \frac{dx_i}{x_i} = \ln|x_i|\right) \tag{8.140}$$

Hence,

$$d \ln x_1 = -\left(\frac{x_2}{x_1}\right) d \ln x_2 \tag{8.141}$$

Based on Eqs. (8.137) and (8.141),

$$d \ln\left(\frac{a_1}{x_1}\right) = -\left(\frac{x_2}{x_1}\right) d \ln\left(\frac{a_2}{x_2}\right) \tag{8.142}$$

and

$$f_i = \frac{a_i}{x_i} \tag{8.143}$$

Thus,

$$d \ln f_1 = -\left(\frac{x_2}{x_1}\right) d \ln f_2 \tag{8.144}$$

By integrating Eq. (8.144) between x_2' and x_2, we obtain

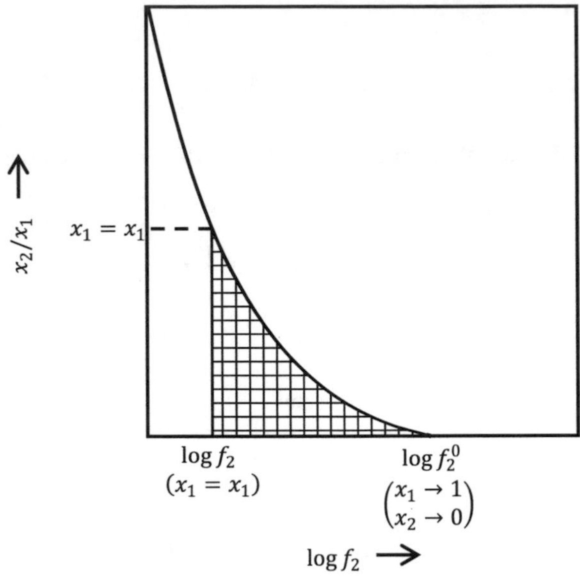

Fig. 8.16 Calculation of activity coefficient using the Gibbs-Duhem equation

$$[\ln f_1]_{x_2'}^{x_2} = -\int_{x_2'}^{x_2} \left(\frac{x_2}{x_1}\right) \mathrm{d} \ln f_2 \tag{8.145}$$

When $x_2' \to 0$, $x_1' \to 1$, $f_1 = 1$, and $f_2 = f_2^0$. Thus, f_1 at concentration $x_1 = 1 - x_2$ is

$$\ln f_1 = -\int_0^{x_2} \left(\frac{x_2}{x_1}\right) \mathrm{d} \ln f_2 \tag{8.146}$$

or

$$\log f_1 = -\int_{\log f_2^0}^{\log f_2} \left(\frac{x_2}{x_1}\right) \mathrm{d} \log f_2 \tag{8.147}$$

Hence, if the activity coefficient f_2 and concentration are known, $\log f_1$ can be calculated using graphical integration (the hatched area in Fig. 8.16).

Example 8.8

Calculate the a_{Fe} value of a Fe–Si alloy at 1873 K (1600 °C) for each concentration using the Gibbs-Duhem equation.

Use the $a_{Si}^{(R)}$ values in Table 8.2.

Solution:

Using the f_{Si} values that can be calculated using the values in Table 8.2$\left(f_{Si} = a_{Si}^{(R)}/x_{Si}\right)$ and $f_{Si}^{0} = 0.0011$ (Fig. 10.2), the relationship between $\log f_{Si}$ and x_{Si}/x_{Fe} can be plotted as in Fig. 8.16. By performing graphical integration, the relationship between x_{Fe} and f_{Fe} can be obtained. We can then obtain a_{Fe} from $a_{Fe} = f_{Fe}x_{Fe}$.
The calculation results are summarised below.

x_{Fe}	$a_{Fe}^{(R)}$
0.1	0.0023
0.2	0.0068
0.3	0.0143
0.4	0.271
0.5	0.0558
0.6	0.142
0.7	0.366
0.8	0.682
0.9	0.955

8.4.3 Calculating Activity Using the Heat of Mixing

If the activity values are not known, we can assume that the solution is an ideal solution and $a_i = x_i$. However, if the heat of mixing$\left(\Delta_{mix}\overline{H}_i = \overline{H}_i - \overline{H}_i^*\right)$ is known,

Table 8.2 $a_{Si}^{(R)}$ values

x_{Si}	$a_{Si}^{(R)}$
0.0	0.000
0.1	0.00030
0.2	0.00190
0.3	0.0122
0.4	0.0713
0.5	0.223
0.6	0.406
0.7	0.581
0.8	0.742
0.9	0.885
1.0	1.000

we can assume that the solution is a regular solution and calculate activity based on it.

Thermodynamically, solutions can be classified as (1) ideal solutions, (2) regular solutions, or (3) real solutions. It can be argued that a regular solution is closer to a real solution than an ideal one; for example, most binary non-electrolyte solutions, such as molten binary alloys, show regular solution behaviour.

For a regular solution, the following equation is true (the derivation of which is given below):

$$\ln f_i = \frac{\Delta_{\text{mix}} \overline{H}_i}{RT} \tag{8.148}$$

where f_i is the activity coefficient, $\Delta_{\text{mix}} \overline{H}_i$ is the heat of mixing, R is the gas constant, and T is the temperature.

Thus, the activity coefficient and activity can be obtained from $\Delta \overline{H}_i$.

In the discussion that follows, the details of ideal and regular solutions will be given, and Eq. (8.148) will be derived. For ideal solutions, the following equation is true, and is proven by Eq. (8.21):

$$\mu_i^{(\text{sln})} = \mu_i^{*,(\ell)} + RT \ln x_i \tag{8.149}$$

where $\mu_i^{(\text{sln})}$ is the chemical potential of component i in the solution, $\mu_i^{*,(\ell)}$ is the chemical potential of pure component i in liquid form (here, it is the standard chemical potential), and x_i is the mole fraction of component i.

The heat of mixing of a regular solution ($\Delta_{\text{mix}} H \neq 0$) differs from that of an ideal solution ($\Delta_{\text{mix}} H = 0$). However, the partial molar entropy, $\overline{S}_i$, of a regular solution is the same as that of an ideal solution. Thus, it can be argued that the atoms/molecules of a regular solution are randomly distributed.

The partial molar entropy of component i, $\overline{S}_i$, can be expressed as follows:

$$\overline{S}_i = \overline{S}_i^* - R \ln x_i \tag{8.150}$$

where $\overline{S}_i^*$ is the partial molar entropy of pure component i.

This equation can be derived from Eq. (7.32) $\left(\left(\frac{\partial \mu_i}{\partial T}\right)_{P,x_i} = -\overline{S}_i\right)$.

According to Eq. (8.150),

$$\Delta_{\text{mix}} \overline{S}_i = \overline{S}_i - \overline{S}_i^* = -R \ln x_i \tag{8.151}$$

Thus,

$$\begin{aligned} RT \ln a_i &= RT \ln f_i + RT \ln x_i \\ &= \mu_i - \mu_i^\circ = \Delta_{\text{mix}} \overline{H}_i - T \Delta_{\text{mix}} \overline{S}_i \\ &= \Delta_{\text{mix}} \overline{H}_i + RT \ln x_i \end{aligned} \tag{8.152}$$

Based on Eq. (8.152), $RT\ \ln f_i + RT\ \ln x_i = \Delta_{\text{mix}}\overline{H}_i + RT\ \ln x_i$. Hence,

$$\Delta_{\text{mix}}\overline{H}_i = RT\ \ln f_i \tag{8.153}$$

(End of the derivation of Eq. (8.148))

8.4.4 Activity of Multicomponent Solutions

Most of the solutions that exist in metallurgical processes are multicomponent systems, and are dilute solutions.

The calculation method proposed by C. Wagner (*Wagner's equation*) is widely accepted for calculating the activity of multicomponent solutions. Consider the activity coefficient of a solute (e.g. Component 2) in a multicomponent dilute solution consisting of Solvent 1—Solute 2—Solute 3—·····. By performing the Taylor series expansion of the activity coefficient f_2 at the point at which the concentration of the solutes is zero, activity coefficient can be expressed by the following equation due to the fact that the solution is dilute (the term for the differential coefficients of second and higher orders are negligible):

$$\ln f_2 = \ln f_2^0 + x_2\frac{\partial\ \ln f_2}{\partial x_2} + x_3\frac{\partial\ \ln f_2}{\partial x_3} + \cdots\cdots \tag{8.154}$$

where f_2^0 is the activity coefficient of Component 2 in an infinite dilute state in a 1–2 binary system, and x_i is the mole fraction of component i.

The interaction parameters $\varepsilon_2^{(2)}$, $\varepsilon_2^{(3)}$ ······ can be defined as follows:

$$\varepsilon_2^{(2)} \equiv \left(\frac{\partial\ \ln f_2}{\partial x_2}\right)_{x_i,\, i\neq 2,1} \tag{8.155}$$

$$\varepsilon_2^{(3)} \equiv \left(\frac{\partial\ \ln f_3}{\partial x_2}\right)_{x_i,\, i\neq 3,1} \tag{8.156}$$

....

These relationships are illustrated in Fig. 8.17 for a 1–2–3 ternary system. Wagner's equation is intended for use with infinite dilute solutions. Therefore, $\varepsilon_2^{(3)}$ is the value at $x_2 \to 0$, and is equal to c/b. Similarly, the interaction parameter can be defined by the Taylor series expansion at $x_2 = \mathbf{a}$ and $x_3 = 0$, although in this case the value is c'/b'.

In practice, it is more convenient to express the activity coefficient with mass%. In such a case, this can be expressed as follows, in place of Eq. (8.154):

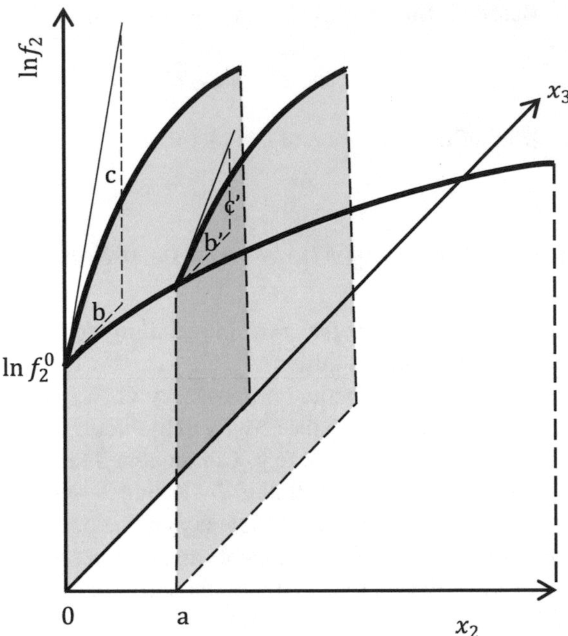

Fig. 8.17 Interaction parameter of 1 (solvent)–2–3 ternary system, $\varepsilon_2^{(3)}$

$$\log \gamma_2 = [\text{mass\% 2}]\frac{\partial \log \gamma_2}{\partial[\text{mass\% 2}]} + [\text{mass\% 3}]\frac{\partial \log \gamma_2}{\partial[\text{mass\% 3}]} + \cdots\cdots \tag{8.157}$$

where the term $\log \gamma_2^0$ is zero as $\gamma_2^0 = 1$, and γ_2 is the activity coefficient of Component 2 (Henrian (1 mass% i) standard state).

The interaction parameters $e_2^{(2)}$, $e_2^{(3)}$ $\cdots\cdots$ can be defined as follows:

$$e_2^{(2)} \equiv \left(\frac{\partial \log \gamma_2}{\partial[\text{mass\% 2}]}\right)_{[\text{mass\%}\ i],\ i\neq 2,1} \tag{8.158}$$

$$e_2^{(3)} \equiv \left(\frac{\partial \log \gamma_2}{\partial[\text{mass\% 3}]}\right)_{[\text{mass\%}\ i],\ i\neq 3,1} \tag{8.159}$$

....

$\varepsilon_i^{(j)}$ and $e_i^{(j)}$ are values relating to infinite dilute solutions. However, it has been experimentally confirmed that these values can be regarded as constant within a certain concentration range. Hence, if the $\varepsilon_i^{(j)}$ and $e_i^{(j)}$ values are known, f_2 and γ_2 at any concentration in the dilute concentration range can be obtained using the following equations:

$$\ln f_2 = \ln f_2^0 + x_2\varepsilon_2^{(2)} + x_3\varepsilon_2^{(3)} + \cdots\cdots \tag{8.160}$$

$$\log \gamma_2 = [\text{mass\% } 2]\mathrm{e}_2^{(2)} + [\text{mass\% } 3]\mathrm{e}_2^{(3)} + \cdots\cdots \tag{8.161}$$

The interaction coefficient of j on i, $f_i^{(j)}$, can be defined as follows:

$$\ln f_2^{(2)} \equiv \ln f_2^0 + x_2\varepsilon_2^{(2)} \tag{8.162}$$

$$\ln f_2^{(3)} \equiv x_3\varepsilon_2^{(3)} \tag{8.163}$$

As their names are similar and thus somewhat confusing, we repeat that $\varepsilon_i^{(j)}$ and $\mathrm{e}_i^{(j)}$ are the interaction parameters and $f_i^{(j)}$ is the interaction coefficient.

The values of $\varepsilon_i^{(j)}$ and $\mathrm{e}_i^{(j)}$ have been experimentally obtained for many alloy systems, with iron as a solvent. The values for $\mathrm{e}_i^{(j)}$ are summarised in Table 13.2 in Sect. 13.2.

The values for $\varepsilon_i^{(j)}$ and $\mathrm{e}_i^{(j)}$ can be mutually converted as follows:

$$\varepsilon_i^{(j)} = 230\left(\frac{M_j}{M_1}\right)\mathrm{e}_i^{(j)} + \frac{(M_1 - M_j)}{M_1} \tag{8.164}$$

Note that

(i) $\varepsilon_i^{(j)} = \varepsilon_j^{(i)}$

(ii) $\mathrm{e}_i^{(j)} \neq \mathrm{e}_j^{(i)}$

(iii) $\mathrm{e}_i^{(j)} = \frac{1}{M_j}\left(M_i\mathrm{e}_j^{(i)} + \frac{M_j - M_i}{230}\right)$

(iv) $\mathrm{e}_i^{(j)} \neq 0$ even if $\varepsilon_i^{(j)} = 0$ (see Sect. 8.2.3)

Example 8.9

Calculate and compare the concentration of nitrogen in rail steel ([mass% C] = 0.6, [mass% Mn] = 0.8, and [mass% Si] = 0.3 mass%) and pig iron ([mass% C] = 4.0, [mass% Mn] = 1.0, and [mass% Si] = 1.0 mass%), which are in equilibrium with 1 bar of N_2 gas at 1873 K (1600 °C).

Use the following ΔG° value for the reaction $\frac{1}{2}N_2(g) = \underline{N}$:

$$\Delta G^\circ = \mu_{\underline{N}}^{\circ(H)} - \frac{1}{2}\mu_{N_2}^{\circ(g)}$$

$$= 3598 + 23.89T \ \mathrm{J \cdot mol^{-1}}$$

Solution:

$$\Delta G^\circ = -RT \ln \frac{a_{\underline{N}}^{(H)}}{P_{N_2}^{1/2}}$$

$$P_{N_2} = 1 \text{ bar}$$

Therefore,

$$3598 + 23.89T = -RT\ \ln a_{\mathrm{N}}^{(\mathrm{H})}$$

$$\log a_{\mathrm{N}}^{(\mathrm{H})} = \frac{1}{2.303}\ln a_{\mathrm{N}}^{(\mathrm{H})} = \frac{3598 + 23.89T}{2.303RT} = -\frac{187.9}{T} - 1.252$$

At $T = 1873$ K (1600 °C),

$$\log a_{\mathrm{N}}^{(\mathrm{H})} = -\frac{187.9}{1873} - 1.252 = -1.351$$

On the other hand,

$$\begin{aligned}\log a_{\mathrm{N}}^{(\mathrm{H})} &= \log(\gamma_{\mathrm{N}}[\text{mass\% N}]) = \log \gamma_{\mathrm{N}} + \log[\text{mass\% N}] \\ &= [\text{mass\% N}]\mathrm{e}_{\mathrm{N}}^{(\mathrm{N})} + [\text{mass\% C}]\mathrm{e}_{\mathrm{N}}^{(\mathrm{C})} + [\text{mass\% Mn}]\mathrm{e}_{\mathrm{N}}^{(\mathrm{Mn})} + [\text{mass\% Si}]\mathrm{e}_{\mathrm{N}}^{(\mathrm{Si})} + \log[\text{mass\% N}]\end{aligned}$$

According to a thermodynamic data book,

$$\mathrm{e}_{\mathrm{N}}^{(\mathrm{N})} = 0,\ \mathrm{e}_{\mathrm{N}}^{(\mathrm{C})} = 0.13,\ \mathrm{e}_{\mathrm{N}}^{(\mathrm{Mn})} = -0.02,\ \mathrm{e}_{\mathrm{N}}^{(\mathrm{Si})} = 0.048$$

By substituting these values into the previous equation, we obtain

$$\log[\text{mass\% N}] = -0.13[\text{mass\% C}] + 0.02[\text{mass\% Mn}] - 0.048[\text{mass\% Si}] - 1.351$$

Therefore,
(1) For rail steel,

$$[\text{mass\% N}] = 0.037\ \text{mass\%}$$

(2) For pig iron,

$$[\text{mass\% N}] = 0.013\ \text{mass\%}$$

Example 8.10
Calculate the activity coefficient of sulphur, γ_{S}, at 1873 K (1600 °C) for rail steel (([mass% C] = 0.6, [mass% Mn] = 0.8, [mass% Si] = 0.3, [mass% S] = 0.03,, and [mass% N] = 0.044 mass%) and pig iron ([mass% C] = 4.0, [mass% Mn] = 1.0, [mass% Si] = 1.0, [mass% S] = 0.03, and [mass% N] = 0.362 mass%).
Solution:

$$\log \gamma_S = [\text{mass\% S}]e_S^{(S)} + [\text{mass\% C}]e_S^{(C)} + [\text{mass\% Mn}]e_S^{(Mn)} + [\text{mass\% Si}]e_S^{(Si)} + [\text{mass\% N}]e_S^{(N)}$$

According to a thermodynamic databook,

$$e_S^{(S)} = 0.046,\ e_S^{(C)} = 0.111,\ e_S^{(Mn)} = -0.026,\ e_S^{(Si)} = 0.075,\ e_S^{(N)} = 0.01$$

By substituting these values into the previous equation, we obtain

$$\log \gamma_S = 0.046[\text{mass\% S}] + 0.111[\text{mass\% C}] - 0.026[\text{mass\% Mn}] + 0.075[\text{mass\% Si}] + 0.01[\text{mass\% N}]$$

Therefore,

(1) For rail steel,

$$\log \gamma_S = 6.73 \times 10^{-3}$$

Hence, $\gamma_S = 1.02$

(2) For pig iron,

$$\log \gamma_S = 0.195$$

Hence, $\gamma_S = 1.57$

8.5 Summary

8.5.1 Chemical Potential and Fugacity/Activity

Chemical potential can be described by the following forms:

$\mu_i = \mu_i^\circ + RT \ln f_i$ for a gas phase

$\mu_i = \mu_i^\circ + RT \ln a_i$ for a condensed phase

Standard chemical potential, μ_i°, and fugacity/activity (f_i or a_i) depend on the system. The standard chemical potential and fugacity/activity terms of each system are summarised in Table 8.3.

The activity of component i is represented by a_i. In metallurgical engineering, the following two conventions are used for standard state and unit of concentration:

Convention (a) (Raoultian standard state)

Standard condition: pure component i

Unit of concentration: mole fraction, x_i

Activity coefficient: f_i.

Chemical potential in a standard state: $\mu_i^{\circ(R)}$.

Table 8.3 Summary of standard chemical potential and fugacity/activity terms

System	Standard chemical potential, μ_i°	Fugacity/Activity term, f_i or a_i
Ideal pure gas	$\mu_i^{\circ,\text{id(g)}}(T, P = 1\text{ bar})$	P
Ideal gas mixture	$\mu_i^{\circ,\text{id(g)}}(T, P = 1\text{ bar})$	$P_i = Px_i$
Ideal solution	$\mu_i^{*,(\ell)}(T, P)$ $\approx \mu_i^{\circ,(\ell)}(T, P = 1\text{ bar})$	x_i
Non-ideal gas	$\mu_i^{\circ,\text{re(g)}}$	f_i
Non-ideal solution (Raoultian standard state)	$\mu_i^{\circ(\text{R})}$	$a_i^{(\text{R})} = f_i x_i$
Non-ideal solution (Henrian standard state)	$\mu_i^{\circ(\text{H})}$	$a_i^{(\text{H})\prime} = \gamma_i' x_i$
Non-ideal solution (Henrian (1 mass% i) standard state)	$\mu_{\underline{i}}^{\circ(\text{H})}$	$a_i^{(\text{H})} = \gamma_i[\text{mass\% } i]$

Convention (b) (Henrian (1 mass% i) Standard state)

Standard condition:

The hypothetical state at [mass% i] = 1 mass%. This can be obtained by extrapolating the linear relationship between the chemical potential (μ_i) and concentration of component i ([mass% i]), which is true for an infinite dilute solution, to [mass% i] = 1 mass% (Fig. 8.7).

Unit of concentration: mass% [mass% i]
Activity coefficient: γ_i.
Chemical potential in a standard state: $\mu_{\underline{i}}^{\circ(\text{H})}$.

Convention (a) is known as the Raoultian standard state as it obeys Raoult's law.

Convention (b) is known as the Henrian standard state as it obeys Henry's law.

In the field of metallurgy, the activity coefficient is conventionally represented by γ_i for the Raoultian standard state and f_i for the Henrian standard state. However, this book uses the symbols recommended by IUPAC; f_i for the Raoultian standard state and γ_i for the Henrian standard state.

8.5.2 *Activity Measurements*

(1) Using electromotive force

$$a_i = e^{-EZ\mathcal{F}/RT} \tag{8.77}$$

where E is the electromotive force, Z is the charge number of ions entering the cell reaction, $\mathcal{F}$ is the Faraday constant $\left(= 9.64853 \times 10^4\ \mathrm{C \cdot mol^{-1}}\right)$, R is the gas constant, and T is the temperature.

(2) Using an oxygen concentration cell consisting of solid electrolyte

$$\Delta G = -4\mathcal{F}E = RT\ \ln\left(\frac{P_{O_2}}{P_{0,O_2}}\right) \tag{8.84}$$

where P_{0,O_2} is the reference partial oxygen pressure.
Partial oxygen pressure, P_{O_2}, can be obtained by measuring electromotive force, E, at temperature T.

(3) Using vapour pressure

$$a_i^{(R)} = f_i x_i = \frac{P_i}{P_i^*} \tag{8.85}$$

where $a_i^{(R)}$ is the activity, f_i is the activity coefficient,x_i is the mole fraction of component i, P_i is the vapour pressure of component i of the solution at concentration x_i, and P_i^* is the vapour pressure of pure component i.

(4) Using the partition constant

$$\left(K_D^\circ\right)_i = \frac{a_{i,1}}{a_{i,2}} \tag{8.93}$$

where $\left(K_D^\circ\right)_i$ is the partition constant, $a_{i,1}$ is the activity of component i in Phase 1, and $a_{i,2}$ is the activity of component i in Phase 2.

(5) Using osmotic pressure

$$RT\ \ln a_1 = -\Pi \overline{V}_1(P_0) \tag{8.100}$$

where a_1 is the activity of Component 1, Π is the osmotic pressure($\Pi = P - P_0$), P and P_0 are the pressure of the solution phase and pure liquid phase, respectively; R is the gas constant, T is the temperature, and $\overline{V}_1(P_0)$ is the partial molar volume of Component 1 in the solution at pressure P_0.

8.5.3 Activity Calculation

(1) Using a phase diagram (activity of a solvent)

$$\ln a_1^{(R)} = -\left(\frac{\Delta_m \overline{H}_1^*}{R}\right)\left(\frac{1}{T} - \frac{1}{T_m}\right) \tag{8.109}$$

where $a_1^{(R)}$ is the activity of Component 1 (Raoultian standard state), $\Delta_m\overline{H}_1^*$ is the molar heat of melting of pure Component 1 at the melting point T_m, R is the gas constant, and T is the temperature.

(2) Using a phase diagram (activity of a solute)

$$\ln a_1 - \ln[\text{mass\% } 1 \text{ at } T_0] = \int_{T_0}^{T} \frac{\Delta\overline{H}_1^{\circ(H)}}{RT^2}\mathrm{d}T \tag{8.120}$$

where $a_1 = \gamma_1[\text{mass\% } 1]$, $\overline{H}_1^{\circ(H)}$ is the partial molar enthalpy of Component 1 in an infinite dilute solution, and [mass% 1 at T_0] is very small, thus $\gamma_1 \approx 1$.

(3) Using the Gibbs-Duhem equation

$$\ln f_1 = -\int_0^{x_2}\left(\frac{x_2}{x_1}\right)\mathrm{d}\ln f_2 \tag{8.146}$$

$$\log f_1 = -\int_{\log f_2^0}^{\log f_2}\left(\frac{x_2}{x_1}\right)\mathrm{d}\log f_2 \tag{8.147}$$

where f_1 is the activity coefficient and x_i is the mole fraction of component i.

(4) Using the heat of mixing
By assuming a regular solution,

$$\ln f_i = \frac{\Delta_{\text{mix}}\overline{H}_i}{RT} \tag{8.148}$$

where f_i is the activity coefficient, $\Delta_{\text{mix}}\overline{H}_i$ is the heat of mixing, R is the gas constant, and T is the temperature.

(5) Activity of multicomponent solutions

$$\ln f_2 = \ln f_2^0 + x_2\varepsilon_2^{(2)} + x_3\varepsilon_2^{(3)} + \cdots\cdots \tag{8.160}$$

$$\log\gamma_2 = [\text{mass\% } 2]\mathrm{e}_2^{(2)} + [\text{mass\% } 3]\mathrm{e}_2^{(3)} + \cdots\cdots \tag{8.161}$$

where f_2 is the activity coefficient (Raoultian standard state), f_2^0 is the activity coefficient of Component 2 when $x_2 \to 0$, x_i is the mole fraction of component i, $\varepsilon_i^{(j)}$ is the interaction parameter, γ_2 is the activity coefficient (Henrian standard state), and $\mathrm{e}_i^{(j)}$ is the interaction parameter.

Table 8.4 Summary of the thermodynamic properties of different types of solution

	$\Delta_{\text{mix}} H$	$\Delta_{\text{mix}} S$
Ideal solution	$\Delta_{\text{mix}} H = 0$	$\Delta_{\text{mix}} S = -R \sum (n_i \ln x_i)$ (see Eq. (6.21) for a binary case)
Regular solution	$\Delta_{\text{mix}} H \neq 0$	$\Delta_{\text{mix}} S = -R \sum (n_i \ln x_i)$ (same as for an ideal solution case)
Real solution	$\Delta_{\text{mix}} H \neq 0$	$\Delta_{\text{mix}} S \neq -R \sum (n_i \ln x_i)$

8.5.4 Types of Solution, Heat of Mixing, $\Delta_{mix} H$, and Entropy of Mixing, $\Delta_{mix} S$

See Table 8.4.

As can be seen in this table, a regularsolution can be defined as a solution in which $\Delta_{\text{mix}} S$ is the same as for an ideal solution, but $\Delta_{\text{mix}} H$ is not.

Reference

Kirkwood JG and Oppenheim I (1961) Chemical thermodynamics, 1st edn. McGraw-Hill

Chapter 9
Partial Molar Quantities and Excess Quantities

In this chapter, several important quantities that are related to extensive properties such as enthalpy, H, entropy, S, and Gibbs energy, G, are presented. The definitions of partial molar quantities, integral quantities, and the excess quantities of enthalpy, entropy, and Gibbs energy will be discussed.

9.1 Partial Molar and Integral Quantities

All extensive properties, including Gibbs energy, enthalpy, entropy, and even volume, can be approached in the same way. As an example, we will examine the partial molar and integral quantities of Gibbs energy, G, which can also be defined for enthalpy, H, and entropy, S. In these cases, Gibbs energy and its symbol, G, can be replaced with H for Enthalpy and S for entropy.

Figure 9.1 shows the Gibbs energy per mole of an A–B binary solution, i.e. the *integral molar Gibbs energy* of a solution, $\overline{G}$.

The integral molar Gibbs energy, $\overline{G}$, of an A–B binary system can be defined by the following equation:

$$\overline{G} = \frac{G}{n_{\mathrm{A}} + n_{\mathrm{B}}} \tag{9.1}$$

where G is the total Gibbs energy of a $n_{\mathrm{A}} + n_{\mathrm{B}}$ mole solution.

The intercepts (at $x_{\mathrm{B}} = 0$ and $x_{\mathrm{B}} = 1$) of the tangent line of the curve $\overline{G}$ at composition x_{B} correspond to the *partial molar Gibbs energy* of Component A, $\overline{G}_{\mathrm{A}}$, and the *partial molar Gibbs energy* of Component B, $\overline{G}_{\mathrm{B}}$, respectively (see Fig. 9.1). Partial molar Gibbs energy can be expressed as follows:

$$\overline{G}_{\mathrm{A}} = \left(\frac{\partial G}{\partial n_{\mathrm{A}}}\right)_{P,T,n_{\mathrm{B}}} \tag{9.2}$$

T. Matsushita and K. Mukai, *Chemical Thermodynamics in Materials Science*,
https://doi.org/10.1007/978-981-13-0405-7_9

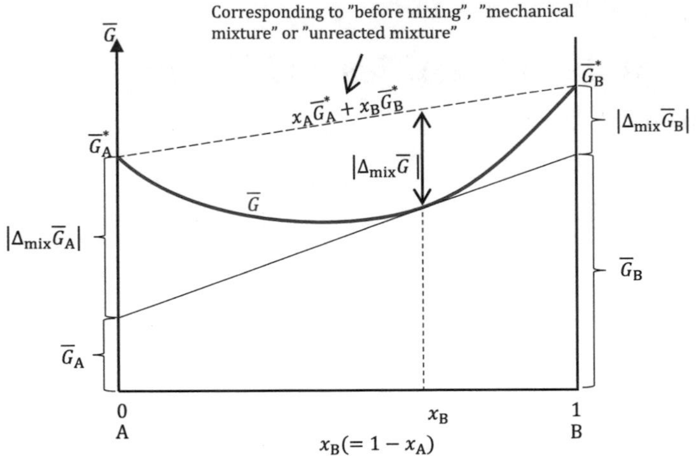

Fig. 9.1 Partial molar and integral quantities

where n_A is the number of moles of Component A.

As can be seen in Eq. (7.59), the partial molar Gibbs energy of component i, $\overline{G}_i$, is identical to the chemical potential of component i, μ_i.

$\overline{G}_i^*$ in Fig. 9.1 is the Gibbs energy of 1 mol of pure component i. The line $x_A\overline{G}_A^* + x_B\overline{G}_B^*$ corresponds to the Gibbs energy of 'before mixing', 'mechanical mixture', or 'unreacted mixture'.

The difference between the line $x_A\overline{G}_A^* + x_B\overline{G}_B^*$ and the integral molar Gibbs energy of solution $\overline{G}$ at composition x_B is termed the *molar Gibbs energy of mixing* (or the *relative integral molar Gibbs energy*), $\Delta_{mix}\overline{G}$, and can be expressed as follows:

$$\Delta_{mix}\overline{G} = \overline{G} - \left(x_A\overline{G}_A^* + x_B\overline{G}_B^*\right) \tag{9.3}$$

For a $n_A + n_B$ mole solution, the *Gibbs energy of mixing*, $\Delta_{mix}G$, can be expressed as follows:

$$\Delta_{mix}G = G - \left(n_A G_A^* + n_B G_B^*\right) \tag{9.4}$$

The integral molar Gibbs energy of solution, $\overline{G}$, can be expressed as follows using partial molar Gibbs energies:

$$\overline{G} = x_A\overline{G}_A + x_B\overline{G}_B \tag{9.5}$$

Therefore, $\Delta_{mix}\overline{G}$ can also be expressed by the following equation:

$$\Delta_{mix}\overline{G} = x_A\overline{G}_A + x_B\overline{G}_B - \left(x_A\overline{G}_A^* + x_B\overline{G}_B^*\right) \tag{9.6}$$

The *relative partial molar Gibbs energy* of component *i* (or the *partial molar Gibbs energy of mixing* of component *i*), $\Delta_{\text{mix}}\overline{G}_i$, can be defined as follows:

$$\Delta_{\text{mix}}\overline{G}_i \equiv \overline{G}_i - \overline{G}_i^* \tag{9.7}$$

In Fig. 9.1, $\Delta_{\text{mix}}\overline{G}_i < 0$

$\Delta_{\text{mix}}\overline{G}_i$ can then be expressed as follows:

$$\Delta_{\text{mix}}\overline{G}_i = \overline{G}_i - \overline{G}_i^* = \left(\mu_i^* + RT\ \ln a_i\right) - \mu_i^* = RT\ \ln a_i \tag{9.8}$$

9.2 Excess Quantities

Thus far, quantities have been defined relative to mechanical mixing. It is, however, also possible to define quantities relative to ideal solutions, allowing excess quantities to be defined.

Figure 9.2 shows molar Gibbs energy of mixing, $\Delta_{\text{mix}}\overline{G}$, against composition.

The difference between the $\Delta_{\text{mix}}\overline{G}$ curve of an ideal solution $\left(\Delta_{\text{mix}}\overline{G}^{\text{id}}\right)$ and that of a non-ideal solution is termed the *excess integral molar Gibbs energy* (or *excess molar Gibbs energy of solution*), $\overline{G}^{\text{E}}$, and can be expressed as follows:

$$\overline{G}^{\text{E}} = \Delta_{\text{mix}}\overline{G} - \Delta_{\text{mix}}\overline{G}^{\text{id}} \tag{9.9}$$

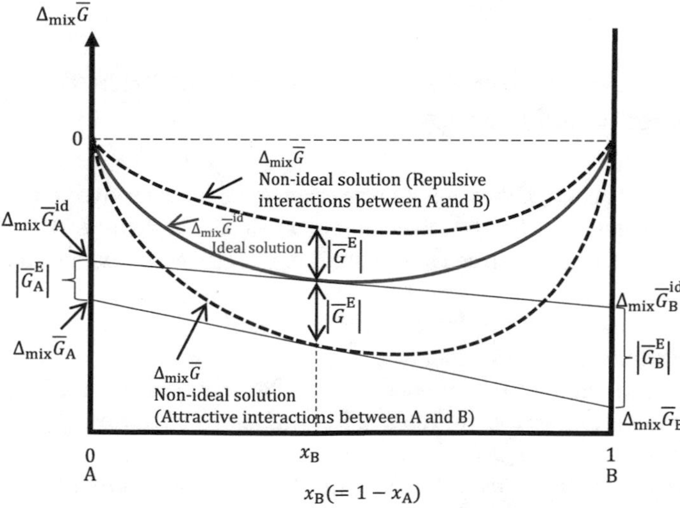

Fig. 9.2 Excess quantities

When a repulsive interaction exists between A and B, $\overline{G}^{\mathrm{E}}$ is positive; when an attractive interaction exists between A and B, $\overline{G}^{\mathrm{E}}$ is negative. Similarly, the *excess partial molar Gibbs energy* of component i (or the *excess relative partial molar Gibbs energy* of component i), $\overline{G}_{\mathrm{i}}^{\mathrm{E}}$, can be defined as follows:

$$\overline{G}_i^{\mathrm{E}} = \Delta_{\mathrm{mix}}\overline{G}_i - \Delta_{\mathrm{mix}}\overline{G}_i^{\mathrm{id}} \tag{9.10}$$

Here,

$$\Delta_{\mathrm{mix}}\overline{G}_i = RT\ \ln a_i \tag{9.11}$$

$$\Delta_{\mathrm{mix}}\overline{G}_i^{\mathrm{id}} = RT\ \ln x_i \tag{9.12}$$

Hence,

$$\overline{G}_i^{\mathrm{E}} = RT\ \ln a_i - RT\ \ln x_i = RT\ \ln f_i \tag{9.13}$$

As can be seen from the above equation, the excess partial molar Gibbs energy of component i, $\overline{G}_i^{\mathrm{E}}$, can be described by the activity coefficient, f_i. We have already seen that the deviation of chemical potential from an ideal solution is $RT\ \ln f_i$ (see e.g. Sect. 8.2.3.1).

The same result can be derived from the difference between the partial molar Gibbs energy of component i of a non-ideal solution, $\overline{G}_i$, and that of an ideal solution, $\overline{G}_i^{\mathrm{id}}$:

$$\overline{G}_i^{\mathrm{E}} = \overline{G}_i - \overline{G}_i^{\mathrm{id}} = \mu_i^\circ + RT\ \ln a_i - \left(\mu_i^\circ + RT\ \ln x_i\right) = RT\ \ln f_i \tag{9.14}$$

9.3 Partial Molar, Integral, and Excess Quantities of Enthalpy and Entropy

In the previous section, Gibbs energy was taken to be an extensive property for the purposes of explanation, but the same quantities can also be defined for enthalpy and entropy. Some of the terms listed below are rarely used, but are defined and listed to provide analogy.

Enthalpy (which can be replaced by the term 'heat'):

Integral molar enthalpy of solution, $\overline{H}$

Partial molar enthalpy of component i, $\overline{H}_i$ (this term is discussed in Sect. 4.4.)

Molar enthalpy of mixing (or *relative integral molar enthalpy*), $\Delta_{\mathrm{mix}}\overline{H}$

Enthalpy of mixing, $\Delta_{\mathrm{mix}}H$

Relative partial molar enthalpy of component i (or *partial molar enthalpy of mixing* of component i), $\Delta_{\mathrm{mix}}\overline{H}_i$

In the case of a regular solution, $\Delta_{\mathrm{mix}}\overline{H}_i = RT\ \ln f_i = \overline{G}_i^{\mathrm{E}}$

Excess integral molar enthalpy (or *excess molar enthalpy of solution*), $\overline{H}^{\mathrm{E}}$

Excess partial molar enthalpy of component i (or *excess relative partial molar enthalpy* of component i), $\overline{H}_i^{\mathrm{E}}$

Integral heat of solution and *differential heat of solution*:

Based on Eq. (9.6), the molar enthalpy of mixing, $\Delta_{\mathrm{mix}}\overline{H}$, can be described as follows:

$$\Delta_{\mathrm{mix}}\overline{H} = x_{\mathrm{A}}\overline{H}_{\mathrm{A}} + x_{\mathrm{B}}\overline{H}_{\mathrm{B}} - \left(x_{\mathrm{A}}\overline{H}_{\mathrm{A}}^{*} + x_{\mathrm{B}}\overline{H}_{\mathrm{B}}^{*}\right) \tag{9.15}$$

where $\overline{H}_i^{*}$ is the enthalpy of 1 mol of pure component i.

The enthalpy of mixing of a $n(= n_{\mathrm{A}} + n_{\mathrm{B}})$ mole solution is

$$\Delta_{\mathrm{mix}}H = n_{\mathrm{A}}\overline{H}_{\mathrm{A}} + n_{\mathrm{B}}\overline{H}_{\mathrm{B}} - \left(n_{\mathrm{A}}\overline{H}_{\mathrm{A}}^{*} + n_{\mathrm{B}}\overline{H}_{\mathrm{B}}^{*}\right) \tag{9.16}$$

where the enthalpy of mixing, $\Delta_{\mathrm{mix}}H$, is also termed the *integral heat of solution.* By differentiating the above equation with respect to n_{A} and n_{B} at constant P and T, we obtain

$$\frac{\partial \Delta_{\mathrm{mix}}H}{\partial n_{\mathrm{A}}} = \overline{H}_{\mathrm{A}} - \overline{H}_{\mathrm{A}}^{*} \tag{9.17}$$

$$\frac{\partial \Delta_{\mathrm{mix}}H}{\partial n_{\mathrm{B}}} = \overline{H}_{\mathrm{B}} - \overline{H}_{\mathrm{B}}^{*} \tag{9.18}$$

As can be seen in Eq. (9.7),

$$\Delta_{\mathrm{mix}}\overline{H}_i \equiv \overline{H}_i - \overline{H}_i^{*} \tag{9.19}$$

Hence,

$$\frac{\partial \Delta_{\mathrm{mix}}H}{\partial n_i} = \Delta_{\mathrm{mix}}\overline{H}_i \tag{9.20}$$

$\Delta_{\mathrm{mix}}\overline{H}_i$, which is identical to the relative partial molar enthalpy of component i, is also termed the *differential heat of solution* of component i. The physicochemical meaning of Eq. (9.20) is that $\Delta_{\mathrm{mix}}\overline{H}_i$ is the heat absorbed (per mole of component i) when a small amount of component i is dissolved in a solution (infinitesimal change in component i) at constant P and T, in other words, the change in heat when 1 mol of component i is dissolved in an infinite amount of solution at constant P and T.

Entropy:

Integral molar entropy of solution, $\overline{S}$

Partial molar entropy of component i, $\overline{S}_i$

Molar entropy of mixing (or *relative integral molar entropy*), $\Delta_{\mathrm{mix}}\overline{S}$

Entropy of mixing, $\Delta_{\mathrm{mix}}S$

Relative partial molar entropy of component i (or *partial molar entropy of mixing* of component i), $\Delta_{\mathrm{mix}}\overline{S}_i$

Excess integral molar entropy (or *excess molar entropy of solution*), $\overline{S}^{\mathrm{E}}$

Excess partial molar entropy of component *i* (or *excess relative partial molar entropy* of component *i*), $\overline{S}_i^{\mathrm{E}}$.

9.4 Calculation of Partial Molar Quantities

In the case of an A–B binary system, the partial molar quantity (for example, the partial molar Gibbs energy of B, $\overline{G}_{\mathrm{B}}$) can be calculated using the following equation (see Fig. 9.3):

$$\overline{G}_{\mathrm{B}} = \overline{G} + (1 - x_{\mathrm{B}})\frac{\partial \overline{G}}{\partial x_{\mathrm{B}}} = \overline{G} + \frac{\partial \overline{G}}{\partial x_{\mathrm{B}}} - x_{\mathrm{B}}\frac{\partial \overline{G}}{\partial x_{\mathrm{B}}} \tag{9.21}$$

Analogically, the partial molar quantities of a multicomponent system (for example, the partial molar Gibbs energy of component i, $\overline{G}_i$) can be calculated using the following equation:

$$\overline{G}_i = \overline{G} + (1 - x_i)\frac{\partial \overline{G}}{\partial x_i} - \sum_{k \neq i} x_k \left(\frac{\partial \overline{G}}{\partial x_k}\right) \tag{9.22}$$

G in the above equations can be replaced by any extensive property, such as enthalpy, H, and entropy, S.

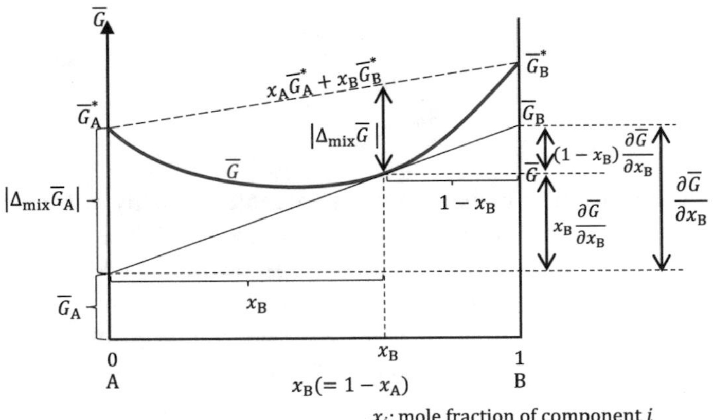

Fig. 9.3 Calculation of partial molar quantities

9.5 Summary

In this chapter, the technical terms relating to extensive properties have been summarised. Among other concepts, partial molar quantities and their calculation methods, integral quantities, properties related to mixing, and the concept of excess quantities, have been discussed, and have been shown to be applicable to any extensive property, including Gibbs energy, enthalpy, entropy, and volume.

Chapter 10
Gibbs Energy Change, ΔG, and Standard Gibbs Energy Change, $\Delta G°$

The previous chapters described theories relating to equilibrium conditions; this chapter discusses utilising these theories in practice. We will first consider the van't Hoff isotherm, which ascertains the direction of a reaction under certain conditions, and will then discuss the Ellingham diagram, which assists in ascertaining the stability of oxides, sulphides, etc. in an easy way.

10.1 The van't Hoff Isotherm

According to Sect. 7.4, the condition for equilibrium for a chemical reaction,

$$\sum_{i=1}^{r} \nu_i \mathrm{X}_i = 0 \tag{10.1}$$

is

$$\Delta G = \sum_{i=1}^{r} \nu_i \mu_i = 0 \tag{10.2}$$

where ΔG is the Gibbs energy change, X_i is 1 mol of a pure chemical substance in a stable state of aggregation at the temperature T and the pressure P at which the reaction occurs, ν_i corresponds to the stoichiometric number (which is negative for reactants and positive for products), and μ_i is the chemical potential (partial molar Gibbs energy).

By substituting the relationship between chemical potential and activity, $\mu_i = \mu_i^\circ + RT\ \ln a_i$, into Eq. (10.2), we obtain

$$\sum_{i=1}^{r} \left[\nu_i \mu_i^\circ + \nu_i RT\ \ln a_i\right] = 0 \tag{10.3}$$

T. Matsushita and K. Mukai, *Chemical Thermodynamics in Materials Science*,
https://doi.org/10.1007/978-981-13-0405-7_10

where μ_i° is the standard chemical potential of component i, R is the gas constant, T is the temperature, and a_i is the activity of component i.

Here, we define

$$\sum_{i=1}^{r} \left(\nu_i \mu_i^\circ\right) \equiv \Delta G^\circ \tag{10.4}$$

and

$$\sum_{i=1}^{r} \left(\nu_i RT \ln a_i\right) \equiv RT \ln K \tag{10.5}$$

where ΔG° is the *standard Gibbs energy change* and K is the equilibrium constant.

Therefore, Eq. (10.3) can be rewritten as

$$\Delta G^\circ = -RT \ln K \tag{10.6}$$

which is called the *van't Hoff isotherm.*

The equilibrium constant, K, of the reaction is described by

$$K = \prod_{i=1}^{r} a_i^{\nu_i} \tag{10.7}$$

where Π (capital pi) denotes the products (i.e. $\prod_{i=1}^{r} a_i^{\nu_i} = a_1^{\nu_1} \cdot a_2^{\nu_2} \cdot a_3^{\nu_3} \ldots a_r^{\nu_r}$).

In the case of a gas, it can be described as follows by assuming an ideal gas:

$$K = \prod_{i=1}^{r} P_i^{\nu_i} \tag{10.8}$$

10.2 The van't Hoff Isotherm for Different Systems

Let us now discuss the van't Hoff isotherm and its meaning in detail, taking the following reaction as an example:

$$\nu_1' \mathrm{X}_1 + \nu_2' \mathrm{X}_2 = \nu_3' \mathrm{X}_3 + \nu_4' \mathrm{X}_4 \tag{10.9}$$

where $\mathrm{X}_i (i = 1, 2, 3, 4)$ is 1 mol of each component and $\nu_i' (i = 1, 2, 3, 4)$ is the number of moles of each component. Unlike the above-mentioned ν_i, ν_i' is positive for both reactants and products. Hence, in the case of the reaction described in Eq. (10.9), the relationship between ν_i and ν_i' is as follows: $\nu_1' = \nu_1$, $\nu_2' = \nu_2$, $\nu_3' = -\nu_3$, and $\nu_4' = -\nu_4$.

For instance, in the case of the reaction

$$2\mathrm{Al} + \frac{3}{2}\mathrm{O}_2 = \mathrm{Al}_2\mathrm{O}_3 \tag{10.10}$$

$$\mathrm{X}_1 = \mathrm{Al},\ \mathrm{X}_2 = \mathrm{O}_2,\ \mathrm{X}_3 = \mathrm{Al}_2\mathrm{O}_3,\ \nu_1' = \nu_1 = 2,\ \nu_2' = \nu_2 = \frac{3}{2} \text{ and } \nu_3' = -\nu_3 = 1$$

Using Eq. (7.133), the Gibbs energy change, dG, due to the progress of the reaction (dξ) can be described as follows using the extent of reaction ξ:

$$\mathrm{d}G = \left(\nu_3'\mu_3 + \nu_4'\mu_4 - \nu_1'\mu_1 - \nu_2'\mu_2\right)\mathrm{d}\xi \tag{10.11}$$

where 'd' denotes a virtual variation.

According to Eq. (10.2),

$$\nu_3'\mu_3 + \nu_4'\mu_4 - \nu_1'\mu_1 - \nu_2'\mu_2 = \Delta G \tag{10.12}$$

Therefore, it can be understood that ΔG corresponds to the Gibbs energy change when d$\xi = 1$, i.e. when ν_3' and ν_4' moles of Components 3 and 4 are produced by the reaction of ν_1' and ν_2' moles of Components 1 and 2.

In an equilibrium state, this Gibbs energy change is zero.

$$\Delta G = 0 \tag{10.13}$$

10.2.1 Liquid Phases (Reaction in a Solution)

In the case of a reaction in a solution, the chemical potential of component i in the solution can be described as

$$\mu_i = \mu_i^\circ + RT\ \ln a_i \tag{10.14}$$

Hence, from Eqs. (10.2) and (10.6), we obtain

$$\begin{aligned}\Delta G &= \nu_3'\mu_3^\circ + \nu_4'\mu_4^\circ - \nu_1'\mu_1^\circ - \nu_2'\mu_2^\circ + RT\ \ln \frac{a_3^{\nu_3'} a_4^{\nu_4'}}{a_1^{\nu_1'} a_2^{\nu_2'}} \\ &= \Delta G^\circ + RT\ \ln K = 0\end{aligned} \tag{10.15}$$

As is discussed above, $\Delta G^\circ = \nu_3'\mu_3^\circ + \nu_4'\mu_4^\circ - \nu_1'\mu_1^\circ - \nu_2'\mu_2^\circ$ is the standard Gibbs energy change of the reaction, and K is the equilibrium constant.

As stated in Sect. 7.1, a process is irreversible and the reaction system is changed spontaneously when $(\mathrm{d}G)_{T,P} < 0$. According to Eqs. (10.11) and (10.12),

$$(\mathrm{d}G)_{T,P} = \Delta G \mathrm{d}\xi \tag{10.16}$$

Therefore, for $d\xi > 0$, i.e. a forward reaction (one that takes place left to right), the reaction occurs spontaneously when

$$\Delta G < 0 \tag{10.17}$$

For $d\xi < 0$, i.e. a backward reaction (one that takes place right to left), the reaction occurs spontaneously when

$$\Delta G > 0 \tag{10.18}$$

As can be seen from the following equation,

$$\Delta G = \Delta G^\circ + RT \ln \frac{a_3^{\nu_3'} a_4^{\nu_4'}}{a_1^{\nu_1'} a_2^{\nu_2'}} = \Delta G^\circ + RT \ln K' \tag{10.19}$$

ΔG can be either positive or negative, depending on the value of $K' \left(\equiv \frac{a_3^{\nu_3'} a_4^{\nu_4'}}{a_1^{\nu_1'} a_2^{\nu_2'}} \right)$ and the state of the system. By substituting $\Delta G^\circ = -RT \ln K$ into the above equation, we obtain

$$\Delta G = RT \ln \frac{K'}{K} \tag{10.20}$$

Hence,

$\Delta G < 0$ when $\frac{K'}{K} < 1$
$\Delta G = 0$ when $\frac{K'}{K} = 1$
$\Delta G > 0$ when $\frac{K'}{K} > 1$

10.2.2 Gas Phases

Consider the following reaction of gases:

$$\nu_1' X_1(g) + \nu_2' X_2(g) = \nu_3' X_3(g) + \nu_4' X_4(g) \tag{10.21}$$

where g denotes the gas phase, $X_i (i = 1, 2, 3, 4)$ is 1 mol of each component, and $\nu_i' (i = 1, 2, 3, 4)$ is the number of moles of each component.

The standard Gibbs energy change, ΔG°, of this reaction is

$$\Delta G^\circ = \nu_3' \mu_3^{\circ,(g)} + \nu_4' \mu_4^{\circ,(g)} - \left(\nu_1' \mu_1^{\circ,(g)} + \nu_2' \mu_2^{\circ,(g)} \right) \tag{10.22}$$

where $\mu_i^{\circ,(g)}$ is the Gibbs energy per mole of component i at a partial pressure of 1 bar, i.e. standard chemical potential.

In this case, the equilibrium constant, K, is

$$K = \frac{P_3^{\nu_3'} P_4^{\nu_4'}}{P_1^{\nu_1'} P_2^{\nu_2'}} \tag{10.23}$$

where P_i is the partial pressure of component i (bar). Strictly speaking, fugacity, f_i, should be used, although up to around 1 bar $P_i = f_i$.

K depends only on temperature, and not on total pressure. The partial pressures on the right-hand side of the chemical equation (products) are the numerators, and the partial pressures on the left-hand side of the chemical equation (reactants) are the denominators. This relationship can be applied to all reaction systems.

Example 10.1

H_2–H_2O gas mixtures, of which an example is shown in the following equation, are important systems for controlling partial oxygen pressure at elevated temperatures:

$$H_2(g) + \frac{1}{2}O_2(g) = H_2O(g)$$

ΔG° for this reaction is

$$\Delta G^\circ = \mu_{H_2O}^{\circ,(g)} - \left(\mu_{H_2}^{\circ,(g)} + \frac{1}{2}\mu_{O_2}^{\circ,(g)}\right)$$
$$= -239530 + 18.74T \log T - 9.247T \text{ J} \cdot \text{mol}^{-1}$$

The equilibrium constant, K, is

$$K = \frac{P_{H_2O}}{P_{H_2} P_{O_2}^{1/2}}$$

Calculate the partial oxygen pressure at 1273 K (1000 °C).

Solution:

By substituting $T = 1000 + 273 = 1273$ K into the equation for ΔG°, $\Delta G^\circ_{1273} = -177220 \text{ J} \cdot \text{mol}^{-1}$.

$$\Delta G^\circ = -RT \ln K$$

where R is the gas constant ($8.314 \text{ J} \cdot \text{mol}^{-1} \cdot \text{K}^{-1}$).

Therefore,

$$K = 1.87 \times 10^7$$

Thus, for example,
$P_{O_2} = 2.86 \times 10^{-15}$bar at $\frac{P_{H_2O}}{P_{H_2}} = 1$
$P_{O_2} = 2.86 \times 10^{-35}$bar at $\frac{P_{H_2O}}{P_{H_2}} = 10^{-10}$

Example 10.2
The following gas mixture is also used to control partial oxygen pressure:

$$CO(g) + \frac{1}{2}O_2(g) = CO_2(g)$$

Find the relationship between log P_{O_2} and log $\frac{P_{CO_2}}{P_{CO}}$ at 1273 K (1000 °C).
$\Delta G°$ for the above reaction is $\Delta G° = -282420 + 86.82T\ \text{J} \cdot \text{mol}^{-1}$.
Solution:

$$K = \frac{P_{CO_2}}{P_{CO} P_{O_2}^{1/2}}$$

$$\log P_{O_2} = 2\log\left(\frac{P_{CO_2}}{P_{CO}}\right) - 2\log K$$

On the other hand, at $T = 1273$ K,

$$\Delta G° = -282420 + 86.82 \times 1273 = 171900\ \text{J} \cdot \text{mol}^{-1}$$

Therefore,

$$\log K = -\frac{\Delta G°}{2.303RT} = -\frac{-171900}{2.303 \times 8.314 \times 1273} = 7.052$$

Hence,

$$\log P_{O_2} = 2\log\left(\frac{P_{CO_2}}{P_{CO}}\right) - 14.10$$

10.2.3 Gas, Solid, and Liquid Phases

10.2.3.1 Pure Solid and Liquid Phases

Consider the following reaction:

$$\nu_1' X_1(g) + \nu_2' X_2(s, \ell) = \nu_3' X_3(s, \ell) \tag{10.24}$$

where g, s, and ℓ denote gas, solid, and liquid, respectively. $X_i (i = 1, 2, 3)$ is 1 mol of each component and $\nu_i' (i = 1, 2, 3)$ is the number of moles of each component.

The standard Gibbs energy change, ΔG°, of this reaction is

$$\Delta G^\circ = \nu_3 \mu_3^\circ - \left(\nu_1 \mu_1^{\circ,(g)} + \nu_2 \mu_2^\circ\right) \tag{10.25}$$

where μ_i° is the Gibbs energy per mole of component i in a standard state, i.e. standard chemical potential. The equilibrium constant, K, of the above reaction is

$$K = \frac{a_3^{\nu_3'}}{P_1^{\nu_1'} a_2^{\nu_2'}} \tag{10.26}$$

where P_i is the partial pressure of component i and a_i is the activity of component i, and has the following relationship with concentration x_i (mole fraction):

$$a_i = f_i x_i \tag{10.27}$$

where f_i is the activity coefficient of component i and x_i is the mole fraction of component i. At pure states, $f_i = 1$ and thus $a_i = 1$, as $x_i = 1$.

If there is no mutual solubility between Components 2 and 3 and they are pure,

$$a_2 = a_3 = 1 \tag{10.28}$$

Hence,

$$K = P_1^{-\nu_1'} \tag{10.29}$$

P_1 is termed *dissociation pressure.*

In most cases, the mutual solubilities between metallic oxides and metals and metallic sulphides and metals are negligible. Hence, the equilibrium constant, K, of a formation reaction can be determined using Eq. (10.29). Dissociation pressure, P_1, can be calculated using a kind of nomograph called the *Ellingham diagram*, which is discussed in Sect. 10.5.

We will discuss this reaction system in more detail, using concrete examples to develop a deeper understanding of the van't Hoff isotherm.

Formation and Decomposition of $Al_2O_3(s)$

The standard Gibbs energy change, ΔG°, of the reaction

$$2\text{Al}(\ell) + \frac{3}{2}\text{O}_2(\text{g}) = \text{Al}_2\text{O}_3(\text{s}) \tag{10.30}$$

is expressed by

$$\Delta G^\circ = \mu^{\circ,(\text{s})}_{\text{Al}_2\text{O}_3} - \left(2\mu^{\circ,(\ell)}_{\text{Al}} + \frac{3}{2}\mu^{\circ,(\text{g})}_{\text{O}_2}\right) \tag{10.31}$$

where μ_i° is the standard chemical potential and (s), (ℓ), and (g) denote the state of component i (solid, liquid, and gas, respectively).

The mutual solubility between $Al(\ell)$ and $Al_2O_3(s)$ is virtually negligible. Hence, the equilibrium constant, K, is

$$K = P_{\text{O}_2}^{-\frac{3}{2}} \tag{10.32}$$

where P_{O_2} is the partial oxygen pressure.

However, if the vapour pressure of Al(g) is lower than the equilibrium vapour pressure between Al(g) and Al(s, ℓ), the following reaction and equilibrium constant must be used for the equilibrium of the reaction system:

$$2\text{Al}(\text{g}) + \frac{3}{2}\text{O}_2(\text{g}) = \text{Al}_2\text{O}_3(\text{s}) \tag{10.33}$$

The ΔG° and K values of this reaction are

$$\Delta G^\circ = \mu^{\circ,(\text{s})}_{\text{Al}_2\text{O}_3} - \left(2\mu^{\circ,(\text{g})}_{\text{Al}} + \frac{3}{2}\mu^{\circ,(\text{g})}_{\text{O}_2}\right) \tag{10.34}$$

$$K = P_{\text{Al}}^{-2} P_{\text{O}_2}^{-\frac{3}{2}} \tag{10.35}$$

Values for $\mu^{\circ,(g)}_{Al}$ in the equation may be found in the thermodynamic data books listed in Sect. 12.2.

Example 10.3

The value of ΔG° of the formation reaction of $Al_2O_3(s)$ is

$$\Delta G^\circ = \mu^{\circ,(\text{s})}_{\text{Al}_2\text{O}_3} - \left(2\mu^{\circ,(\ell)}_{\text{Al}} + \frac{3}{2}\mu^{\circ,(\text{g})}_{\text{O}_2}\right)$$

According to a thermodynamic data book,

$$\Delta G^\circ = -1697700 - 15.69T\ \log T + 385.8T\ \text{J}\cdot\text{mol}^{-1}$$

Derive the relationship between dissociation pressure, P_{O_2}, and temperature.
Solution:

$$K = P_{O_2}^{-\frac{3}{2}}$$

Therefore,

$$\log K = -\frac{3}{2}\log P_{O_2} \tag{1}$$

On the other hand,

$$\log K = -\frac{\Delta G^\circ}{2.303RT} = -\frac{-1697700 - 15.69T\ \log T + 385.8T}{2.303\times 8.314T} = \frac{88666}{T} + 0.819\log T - 20.157 \tag{2}$$

By substituting Eq. (2) into Eq. (1), we obtain

$$\log P_{O_2} = -\frac{59111}{T} - 0.546\log T + 13.438 \tag{3}$$

For example, at 1273 K (1000 °C) and 1773 K (1500 °C), P_{O_2} is 2.0×10^{-35} bar and 2.1×10^{-22} bar, respectively, both of which are relatively low.

Formation and Decomposition of $CaCO_3(s)$

The standard Gibbs energy change, ΔG°, and equilibrium constant, K, of the reaction

$$CaO(s) + CO_2(g) = CaCO_3(s) \tag{10.36}$$

are

$$\Delta G^\circ = -168410 + 143.9T\ \text{J}\cdot\text{mol}^{-1} \tag{10.37}$$

and

$$K = P_{CO_2}^{-1} \tag{10.38}$$

The partial pressure of CO_2, P_{CO_2}, in a container is not higher than 1 bar when this reaction occurs at 1 bar (atmospheric pressure). Therefore, as is shown in Fig. 10.1,

(a) If the dissociation pressure, $P_{CO_{2,e}}$, > 1 bar

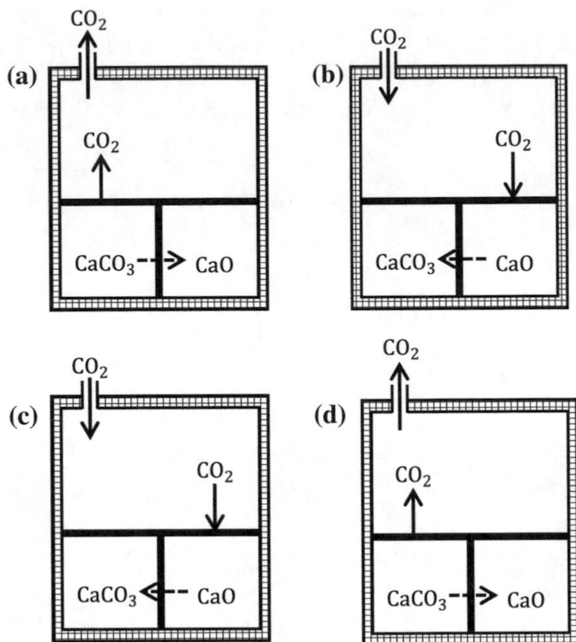

Fig. 10.1 Decomposition and formation of $CaCO_3$. *Note* **a** $P_{CO_2}(= 1\text{bar}) > P_{CO_{2,e}}$, $\Delta G^\circ > 0$. The partial pressure of CO_2, P_{CO_2}, in the container is approaching the dissociation pressure, $P_{CO_{2,e}}$. Therefore, the $CaCO_3$ decomposition reaction will occur. However, if the partial pressure of CO_2, P_{CO_2}, is maintained at 1 bar, the $CaCO_3$ decomposition reaction will continue until all $CaCO_3$ is consumed. Hence, CaO is the stable phase under these conditions. **b** $P_{CO_2}(= 1\text{bar}) > P_{CO_{2,e}}$, $\Delta G^\circ < 0$. The partial pressure of CO_2, P_{CO_2}, in the container is approaching the dissociation pressure, $P_{CO_{2,e}}$. Therefore, $CaCO_3$ formation will occur. However, if the partial pressure of CO_2, P_{CO_2}, is maintained at 1 bar, $CaCO_3$ formation will continue until all CaO is consumed. Hence, $CaCO_3$ is the stable phase under these conditions. **c** $1\text{bar} < P_{CO_{2,e}} < P_{CO_2}(= 4\text{bar})$, $\Delta G^\circ > 0$. P_{CO_2} is approaching $P_{CO_{2,e}}$. Therefore, as with **b**, the $CaCO_3$ formation will continue until all CaO is consumed. Hence, even if $\Delta G^\circ > 0$, $CaCO_3$ is the stable phase (due to $P_{CO_{2,e}} > 1\text{bar}$).**d** $P_{CO_2} < P_{CO_{2,e}} = 1\text{bar})$, $\Delta G^\circ < 0$. P_{CO_2} is approaching $P_{CO_{2,e}}$. Therefore, as with **a**, $CaCO_3$ decomposition will continue until all $CaCO_3$ is consumed. Hence, even if $\Delta G^\circ > 0$, CaO is the stable phase (due to $P_{CO_{2,e}} < 1\text{bar}$)

The partial pressure of CO_2 in a container is always lower than the dissociation pressure, and so the reaction system produces CO_2 as it progresses towards equilibrium, i.e. the decomposition reaction of $CaCO_3$ continues until all $CaCO_3$ is consumed.

(b) If the dissociation pressure, $P_{CO_{2,e}}$, < 1 bar

If the partial pressure of CO_2 in a container is maintained at 1 bar (e.g. by flowing CO_2 gas), the reaction system consumes CO_2 as it progresses towards equilibrium and a forward reaction continues until all CaO is consumed and the $CaCO_3$ is stable.

In the case of (a), $\Delta G^\circ = -RT \ln P_{CO_{2,e}} > 0$ and the decomposition of $CaCO_3$ occurs spontaneously. In the case of (b), $\Delta G^\circ < 0$ and $CaCO_3$ is stable. Therefore, it *seems* that it is possible to estimate the direction of the reaction based on the sign of ΔG°. However, if $P_{CO_2} = 4$ bar and the dissociation pressure is e.g. 2 bar, $CaCO_3$ is stable and the decomposition of $CaCO_3$ does not occur, even though $\Delta G^\circ > 0$ (as shown in Fig. 10.1c). In the scenario outlined in Fig. 10.1d, $CaCO_3$ is not stable even though $\Delta G^\circ < 0$. As we can see from these examples, assuming the direction of a reaction based on the sign of ΔG° is incorrect.

Boudouard Reaction

Consider the following reaction:

$$\mathrm{C(s)} + \mathrm{CO_2(g)} = 2\mathrm{CO(g)} \tag{10.39}$$

This reaction is called the *Boudouard reaction*, and is important in iron and steel-making processes.

The standard Gibbs energy change, ΔG°, and equilibrium constant, K, of this reaction are

$$\Delta G^\circ = 170710 - 174.5T \ \mathrm{J \cdot mol^{-1}} \tag{10.40}$$

and

$$K = \frac{P_{CO}^2}{P_{CO_2}} \tag{10.41}$$

where P_i is the partial pressure of component i.

Equation (10.41) tells us only that $\frac{P_{CO}^2}{P_{CO_2}} < 1$ when $\Delta G^\circ > 0$, and $\frac{P_{CO}^2}{P_{CO_2}} > 1$ when $\Delta G^\circ < 0$. Depending on P_{CO} and P_{CO_2} at a given temperature, the reaction occurs towards the right-hand side even if $\Delta G^\circ > 0$ and towards the left-hand side even if $\Delta G^\circ < 0$.

As can be seen from these examples, the direction of a reaction must be estimated using the sign of ΔG and not the sign of ΔG°. When $\Delta G < 0$, the reaction occurs spontaneously from left to right; when $\Delta G > 0$, the reaction occurs spontaneously from right to left.

Example 10.4

The value of ΔG° for the decomposition reaction of Si_3N_4,

$$\mathrm{Si_3N_4(s)} = 3\mathrm{Si}(\ell) + 2\mathrm{N_2(g)}$$

is

$$\Delta G^\circ = 874460 - 405.0T \text{ J} \cdot \text{mol}^{-1}$$

Determine whether Si_3N_4 is stable under the condition $P_{Ar} + P_{N_2} = 1$ bar, $P_{N_2} = 0.8$ bar, 1973 K (1700 °C).
Solution:

$$\begin{aligned}\Delta G &= \Delta G^\circ + RT \ln P_{N_2}^2 \\ &= 874460 - 405.0T + 8.314T \ln 0.8^2 \\ &= 874460 - 408.7T \text{ J} \cdot \text{mol}^{-1}\end{aligned}$$

At 1973 K (1700 °C),

$$\Delta G = 68074 \text{ J} \cdot \text{mol}^{-1} > 0$$

Hence, under these conditions, Si_3N_4 is stable.

Example 10.5
Thermodynamically, marble is formed through the solidification of molten $CaCO_3$.
Calculate the P_{CO_2} range required to form marble.
The melting point of $CaCO_3$ is 1503 K (1230 °C).
Solution:
The decomposition of $CaCO_3(s)$ is described as follows:

$$CaCO_3(s) = CaO(s) + CO_2(g)$$

ΔG° of this reaction is

$$\Delta G^\circ = 168410 - 143.9T \text{ J} \cdot \text{mol}^{-1}$$

Therefore,

$$\Delta G = \Delta G^\circ + RT \ln P_{CO_2} = 168410 - 143.9T + 8.314T \ln P_{CO_2}$$

If $\Delta G > 0$ at $T > 1503$ K (1230 °C), $CaCO_3(s)$ will melt without decomposition.
Therefore, if $\ln P_{CO_2} > 3.830$ (i.e. $P_{CO_2} > 46.1$ bar), decomposition will not occur and marble will, according to the calculation,be formed.

10.2.3.2 One or Both of the Solid and Liquid Phases Are Solutions

Consider the following equation:

$$\nu_1' X_1(g) + \nu_2' X_2(s, \ell) = \nu_3' X_3(s, \ell) \tag{10.42}$$

where $X_i (i = 1, 2, 3)$ is 1 mol of each component and $\nu_i' (i = 1, 2, 3)$ is the number of moles of each component.

The standard Gibbs energy change, $\Delta G°$, and equilibrium constant, K, of the reaction are

$$\Delta G° = \nu_3' \mu_3^\circ - \left(\nu_1' \mu_1^{\circ,(g)} + \nu_2' \mu_2^\circ\right) \tag{10.43}$$

and

$$K = \frac{a_3^{\nu_3'}}{P_1^{\nu_1'} a_2^{\nu_2'}} \tag{10.44}$$

where P_i is the partial pressure of component i and a_i is the activity of component i.

If there is mutual solubility between Components 2 and 3, the activity of neither Component 2 (a_2) nor Component 3(a_3) is equal to 1.

In the case of an ideal solution, the activity is equal to the concentration, i.e. $a_i = x_i$. Thus,

$$K = \frac{x_3^{\nu_3'}}{P_1^{\nu_1'} x_2^{\nu_2'}} \tag{10.45}$$

where x_i is the mole fraction of component i.

However, an ideal solution in the strict sense of the term is rarely found in real aqueous solutions, molten alloys, slag systems, etc. As an example, the a_{Si} and a_{Fe} values of a Fe–Si alloy system are shown in Fig. 10.2. At a low Si concentration, the value of f_{Si} is roughly 1/1000; therefore, the value of the mole fraction, x_{Si}, is 1000 times larger than the value of the activity, a_{Si}.

As is discussed in Sect. 8.2.2, activity can be determined only after a standard state has been specified, and the activity coefficient can be determined only after the standard state and unit of concentration have been specified. Therefore, the factors of $\Delta G°$ vary with the choice of standard state. In the following examples, these factors, along with the relationship between activity and equilibrium vapour pressure, are discussed.

(i) Reduction of SiO_2 by carbon

CO bubbles are generated by the following reaction, in which high SiO_2 slag comes into contact with pig iron.

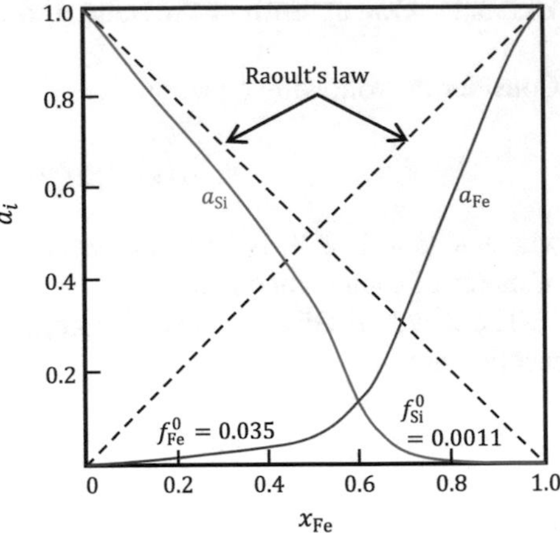

Fig. 10.2 Activity of molten Fe–Si alloy at 1873 K (1600 °C)

$$(SiO_2) + 2\underline{C} = 2CO(g) + \underline{Si} \tag{10.46}$$

The brackets around SiO_2 denote that the component dissolves into the slag, and the underlining denotes that the component dissolves into the metal.

The equilibrium constant, K, of the above reaction is

$$K = \frac{P^2_{CO} a_{Si}}{a_{SiO_2} a^2_C} \tag{10.47}$$

By choosing standard states for a_{Si}, a_{SiO_2}, and a_C (pure liquid Si for a_{Si}, pure solid SiO_2 for a_{SiO_2}, and pure solid C for a_C), ΔG° of the reaction described in Eq. (10.46) is expressed by

$$\Delta G^\circ = 2\mu^{\circ,(g)}_{CO} + \mu^{\circ,(\ell)}_{Si} - \left(\mu^{\circ,(s)}_{SiO_2} + 2\mu^{\circ,(s)}_{C}\right) \tag{10.48}$$

Even if molten iron does not exist in the system, i.e. for the reaction

$$SiO_2(s) + 2C(s) = 2CO(g) + Si(\ell) \tag{10.49}$$

ΔG° can be expressed in a similar manner as follows:

$$\Delta G^\circ = 2\mu^{\circ,(g)}_{CO} + \mu^{\circ,(\ell)}_{Si} - \left(\mu^{\circ,(s)}_{SiO_2} + 2\mu^{\circ,(s)}_{C}\right) \tag{10.50}$$

Therefore, once the value for ΔG° is obtained from the data book, the equilibrium constant, $K = \frac{P^2_{CO} a_{Si}}{a_{SiO_2} a^2_C}$, can be calculated using the van't Hoff isotherm (Eq. (10.6)).

(ii) Equilibrium vapour pressure of each component of an A–B binary solution

Evaporation can be regarded as a reaction in a broad sense, and can be expressed as follows:

$$\underline{\mathrm{A}} = \mathrm{A(g)},\ \Delta G_1^\circ \tag{10.51}$$

$$\underline{\mathrm{B}} = \mathrm{B(g)},\ \Delta G_2^\circ \tag{10.52}$$

$\underline{\mathrm{A}}$ and $\underline{\mathrm{B}}$ denote Components A and B in the solution.

$$\Delta G_1^\circ = \mu_{\mathrm{A}}^{\circ,(\mathrm{g})} - \mu_{\mathrm{A}}^\circ \tag{10.53}$$

$$\Delta G_2^\circ = \mu_{\mathrm{B}}^{\circ,(\mathrm{g})} - \mu_{\mathrm{B}}^\circ \tag{10.54}$$

For Component A,

$$\Delta G_1^\circ = -RT\ \ln \frac{P_{\mathrm{A}}}{a_{\mathrm{A}}} \tag{10.55}$$

At $x_{\mathrm{A}} = 1$,

$$\Delta G_1^\circ = -RT\ \ln P_{\mathrm{A}}^* \tag{10.56}$$

where P_{A}^* is the equilibrium vapour pressure of pure Component A in liquid form.
Therefore,

$$\frac{P_{\mathrm{A}}}{a_{\mathrm{A}}} = P_{\mathrm{A}}^* \tag{10.57}$$

$$P_{\mathrm{A}} = P_{\mathrm{A}}^* f_{\mathrm{A}} x_{\mathrm{A}} \tag{10.58}$$

Similarly,

$$P_{\mathrm{B}} = P_{\mathrm{B}}^* f_{\mathrm{B}} x_{\mathrm{B}} \tag{10.59}$$

A relationship between the composition, total pressure, P, and partial pressures, P_{A} and P_{B}—obtained using the above derivation—is shown in Fig. 10.3 (where $f_{\mathrm{A}} < 1$ and $f_{\mathrm{B}} < 1$).

Convention (b) of Sect. 8.2.3.2 is often used to treat the trace components, i.e. solute components, of a metal. Several examples of this are given in Sect. 10.2.4.2.

10.2.4 Solid and Liquid Phases

The standard Gibbs energy change, ΔG°, and the equilibrium constant, K, of the reaction

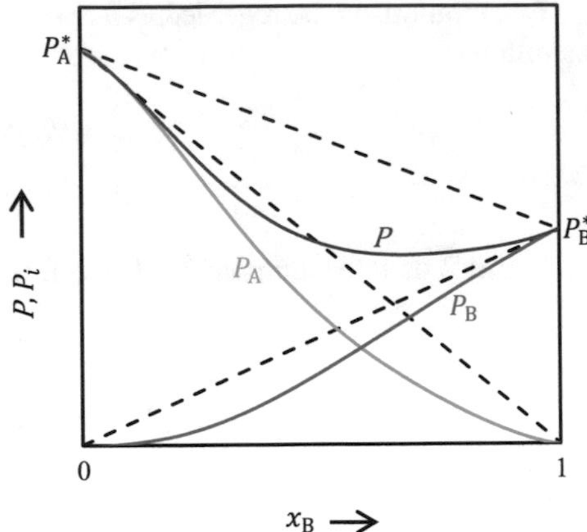

Fig. 10.3 The relationship between composition, total pressure, P, and partial pressure, P_A and P_B

$$\nu_1' X_1(s, \ell) + \nu_2' X_2(s, \ell) = \nu_3' X_3(s, \ell) + \nu_4' X_4(s, \ell) \quad (10.60)$$

are

$$\Delta G^\circ = \nu_3' \mu_3^\circ + \nu_4' \mu_4^\circ - \left(\nu_1' \mu_1^\circ + \nu_2' \mu_2^\circ\right) \quad (10.61)$$

and

$$K = \frac{a_3^{\nu_3'} a_4^{\nu_4'}}{a_1^{\nu_1'} a_2^{\nu_2'}} \quad (10.62)$$

$X_i (i = 1, 2, 3, 4)$ is 1 mol of each component, $\nu_i' (i = 1, 2, 3, 4)$ is the number of moles of each component, μ_i° is the standard chemical potential of component i, and a_i is the activity of component i.

10.2.4.1 Pure Solid and Pure Liquid Phases

In the case of pure solid and pure liquid phases, the activity of each component is equal to 1. Thus,

$$K = 1, \ \Delta G^\circ = 0 \quad (10.63)$$

Therefore, the temperature at which Components 1, 2, 3, and 4 coexist will be defined.

Consider a pure Fe(s)–pure FeO(s)–pure Fe_3O_4(s) coexisting system.

The reaction can be expressed as

$$Fe(s) + Fe_3O_4(s) = 4FeO(s) \quad (10.64)$$

and ΔG° of this reaction is

$$\Delta G^\circ = -52593 + 62.55T\ \mathrm{J \cdot mol^{-1}} \tag{10.65}$$

At 841 K (568 °C), ΔG° is 0.

This temperature corresponds to the coexistence temperature of the above three components in the Fe-O phase diagram, 833 K (560 °C).

10.2.4.2 One or Both of the Solid and Liquid Phases are Solutions

When one or both of the solid and liquid phases are solutions, the activity values of at least one of the components is not equal to 1. This implies that we must ascertain the activity value that is not equal to 1 to ascertain the equilibrium position.

Example 10.6

The reaction between molten iron and a refractory material, SiC, is

$$\underline{\mathrm{Si}} + \underline{\mathrm{C}} = \mathrm{SiC(s)}$$

where $\underline{\mathrm{Si}}$ and $\underline{\mathrm{C}}$ denote dissolved silicon and carbon, respectively.

When ΔG° is expressed as follows:

$$\Delta G^\circ = \mu_{\mathrm{SiC}}^{\circ,(\mathrm{s})} - \left(\mu_{\mathrm{Si}}^{\circ,(\ell)} + \mu_{\mathrm{C}}^{\circ,(\mathrm{s})}\right)$$

it is the same as ΔG° of the following reaction:

$$\mathrm{Si}(\ell) + \mathrm{C(s)} = \mathrm{SiC(s)}$$

According to a data book,

$$\Delta G^\circ = -113390 - 11.42T\ \log T + 75.73T\ \mathrm{J \cdot mol^{-1}}$$

Hence, $\Delta G^\circ_{1773} = -44911\ \mathrm{J \cdot mol^{-1}}$ at 1773 K (1500 °C).

Thus,

$$K = \frac{1}{a_{\mathrm{Si}} a_{\mathrm{C}}} = 21.0$$

When molten iron is saturated with carbon, the activity of carbon, a_{C}, becomes 1 due to the fact that the pure solid carbon, $\mu_{\mathrm{C}}^{\circ,(\mathrm{s})}$, is taken to be a standard state. Therefore, using the above equation, $a_{\mathrm{Si}} = 0.0475$ can be obtained. According to experimental results, saturated carbon concentrations in iron are very small with such a_{Si} values.

From the a_{Si} value shown in Fig. 10.2, the concentration of Si ([mass%]) can be read as [mass%] ≈ 22 mass%. As is shown here, the concentration of Si in the molten iron is determined when the Fe–Si–C system of molten iron, solid carbon, and SiC(s) coexist at a specific temperature. The concentration of carbon is also determined, as the molten iron is saturated by the carbon, and so the composition of the molten iron is determined.

Example 10.7

The deoxidation of molten iron with Si is

$$\underline{Si} + 2\underline{O} = SiO_2(s)$$

the ΔG° and K values of which are expressed as follows:

$$\Delta G^\circ = \mu^{\circ,(s)}_{SiO_2} - \left(\mu^{\circ(H)}_{\underline{Si}} + \mu^{\circ(H)}_{\underline{O}}\right)$$

$$K = \frac{a_{SiO_2}}{a_{Si}a_O^2} = \frac{f_{SiO_2}x_{SiO_2}}{\gamma_{Si}[\text{mass\%Si}]\gamma_O^2[\text{mass\%O}]^2} = \frac{1}{\gamma_{Si}[\text{mass\%Si}]\gamma_O^2[\text{mass\%O}]^2}$$

The value of ΔG° is given by

$$\Delta G^\circ = -588020 + 225.1T \text{ J} \cdot \text{mol}^{-1}$$

According to experimental data, $\gamma_{Si}\gamma_O^2 \approx 1$ at [mass%Si] < 0.5 mass%
Thus,

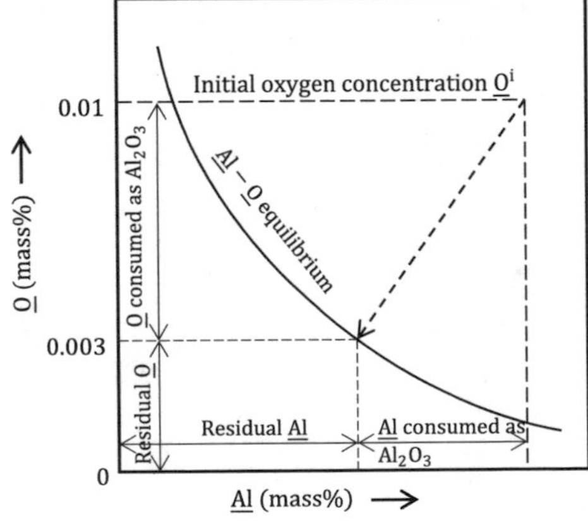

Fig. 10.4 Deoxidation of molten iron by Al

$$K = \frac{1}{[\text{mass\%Si}][\text{mass\%O}]^2}$$

$K = 4.37 \times 10^4$ at 1873 K (1600 °C).
For example, [mass%O] = 0.0106 mass% at [mass%Si] = 0.2 mass%.

Example 10.8

Consider the following reaction:

$$2\underline{\text{Al}} + 3\underline{\text{O}} = \text{Al}_2\text{O}_3(\text{s})$$

ΔG° of this reaction is

$$\Delta G^\circ = \mu^{\circ,(\text{s})}_{\text{Al}_2\text{O}_3} - \left(2\mu^{\circ(\text{H})}_{\underline{\text{Al}}} + 3\mu^{\circ(\text{H})}_{\underline{\text{O}}}\right) = -1242200 + 395.0T\ \text{J}\cdot\text{mol}^{-1}$$

$\gamma^2_{\text{Al}}\gamma^3_{\text{O}} \approx 1$ at [mass%Al] $\leq$ 0.05 mass% (relative error = 30%)

(1) Calculate the solubility product $[\text{mass\%Al}]^2[\text{mass\%O}]^3$ at 1873 K (1600 °C).

(2) Calculate the amount of Al required to decrease the concentration of dissolved oxygen in 1 tonne of iron at 1873 K (1600 °C) from 0.01 mass% to less than 0.003 mass%. As shown in Fig. 10.4, in addition to the Al that is consumed to produce Al_2O_3, the amount of residual Al that is required to suppress the concentration of $\underline{\text{O}}$ to less than 0.003 mass% must be considered.

Solutions:

$$K = \frac{a_{\text{Al}_2\text{O}_3}}{a^2_{\text{Al}}a^3_{\text{O}}} = \frac{1}{\gamma^2_{\text{Al}}[\text{mass\%Al}]^2\gamma^3_{\text{O}}[\text{mass\%O}]^3} = \frac{1}{[\text{mass\%Al}]^2[\text{mass\%O}]^3} \quad (1)$$

At T = 1873 K (1600 °C)

$$\Delta G^\circ = -502370\ \text{J}\cdot\text{mol}^{-1}$$

$$K = 1.02 \times 10^{14}$$

Hence,

$$[\text{mass\%Al}]^2[\text{mass\%O}]^3 = \frac{1}{K} = 9.75 \times 10^{15} \quad (2)$$

Based on the above,
At [mass%O] = 0.01 mass%, [mass%Al] = 0.987×10^{-4} mass%.
At [mass%O] = 0.003 mass%, [mass%Al] = 6.008×10^{-4} mass%.
Therefore, the amount of Al required to suppress the remaining dissolved oxygen in the 1 tonne (1×10^6 g) of molten iron to 0.003 mass% $\underline{\text{O}}$ is

$$1 \times 10^6 \times \left(6.008 \times 10^{-4} - 0.987 \times 10^{-4}\right) \times 10^{-2} = 5.02\,\text{g}$$

The amount of Al that is consumed to form Al_2O_3 is

$$\left[2 \times 27 \times 1 \times 10^6 \times (0.01 - 0.003) \times 10^{-2}\right]/(3 \times 16) = 78.8\,\text{g}$$

Hence, at least 83.8 g (5.02 g + 78.8 g) of Al must be added to 1 tonne of molten iron in order to decrease the concentration of dissolved oxygen by the required amount.

Example 10.9

CO gas generation, which occurs during steelmaking processes, takes place through the following reaction between carbon and oxygen in molten iron:

$$\underline{\text{C}} + \underline{\text{O}} = \text{CO(g)}$$

ΔG° of the reaction is

$$\Delta G^\circ = \mu_{\text{CO}}^{\circ,(g)} - \left(\mu_{\underline{\text{C}}}^{\circ(\text{H})} + \mu_{\underline{\text{O}}}^{\circ(\text{H})}\right) = -22175 - 38.37T\ \text{J} \cdot \text{mol}^{-1}$$

Calculate the [mass%O] for an equilibrium state under the following conditions:
(1) 1823 K (1550 °C), [mass%C] = 0.2 mass%, $P_{\text{CO}} = 1$ bar
(2) 1823 K (1550 °C), [mass%C] = 0.2 mass%, $P_{\text{CO}} = 10^{-3}$ bar
Assume that $\gamma_{\text{O}}\gamma_{\text{C}} \approx 1$.
Solution:
At $T = 1823$ K,

$$\Delta G^\circ = -92123\,\text{J} \cdot \text{mol}^{-1}$$

$$\ln K = 6.078$$

Therefore,

$$K = 436.2 \tag{1}$$

On the other hand,

$$K = \frac{P_{CO}}{a_C a_O} = \frac{P_{CO}}{\gamma_C[\text{mass\%C}]\gamma_O[\text{mass\%O}]} = \frac{P_{CO}}{[\text{mass\%C}][\text{mass\%O}]} = \frac{P_{CO}}{0.2[\text{mass\%O}]} \tag{2}$$

From Eqs. (1) and (2),
At $P_{CO} = 1$ bar, [mass%O] = 0.011 mass%
At $P_{CO} = 10^{-3}$ bar, [mass%O] = 1.1×10^{-5} mass%

Example 10.10
Calculate the solubility of nitrogen in molten iron at $P_{N_2} = 1$ bar and 1823 K (1550 °C), 1873 K (1600 °C), and 1973 K (1700 °C).
ΔG° of the reaction $\frac{1}{2}N_2(g) = \underline{N}$ is

$$\Delta G^\circ = \mu_{\underline{N}}^{\circ(H)} + \frac{1}{2}\mu_{N_2}^{\circ,(g)} = 3598 + 23.89T \text{ J} \cdot \text{mol}^{-1}$$

Assume that $\gamma_N \approx 1$.
Solution:

$$K = \frac{a_N}{P_{N_2}^{1/2}} = \frac{\gamma_N[\text{mass\%N}]}{P_{N_2}^{1/2}} = [\text{mass\%N}]$$

$$\Delta G^\circ = -RT \ln K = -RT \ln[\text{mass\%N}] = 3598 + 23.89T \text{ J} \cdot \text{mol}^{-1}$$

Therefore,

$$[\text{mass\%N}] = \exp\left(-\frac{3598 + 23.89T}{RT}\right)$$

Hence,
At $T = 1823$ K, [mass%N] = 0.044 mass%
At $T = 1873$ K, [mass%N] = 0.045 mass%
At $T = 1973$ K, [mass%N] = 0.045 mass%

Example 10.11

Calculate the solubility of nitrogen in molten iron ([mass%Al] = 1 mass% and [mass%Si] = 2 mass%) at $P_{N_2} = 1$ bar and 1823 K (1550 °C), 1873 K (1600 °C), and 1973 K (1700 °C) using activity coefficients.

Compare these values with the results of calculations that assume that $\gamma_i = 1$ (Example 10.10).

Solution:

$$P_{N_2} = 1\,\text{bar}$$

Therefore, the equilibrium constant, K, is

$$K = a_N / P_{N_2}^{1/2} = a_N = \gamma_N[\text{mass\%N}]$$

As can be seen in this equation, the solubility of a diatomic gas is proportional to the square root of its partial pressure. This relationship is known as *Sieverts' law*.

According to Eq. (8.161),

$$\begin{aligned}\log K &= \log \gamma_N + \log[\text{mass\%N}] \\ &= [\text{mass\%N}]e_N^{(N)} + [\text{mass\%Al}]e_N^{(Al)} \\ &\quad + [\text{mass\%Si}]e_N^{(Si)} + \log[\text{mass\%N}]\end{aligned}$$

According to a thermodynamic data book,
$e_N^{(N)} = 0$, $e_N^{(Al)} = 0.01$, $e_N^{(Si)} = 0.048$
If $e_i^{(j)}$ does not depend on temperature,

$$\log[\text{mass\%N}] = \log K - 0.199$$

On the other hand,

$$\Delta G^\circ = -RT\ \ln K$$

Therefore,

$$\log K = \frac{\ln K}{2.303} = \frac{\Delta G^\circ}{-2.303RT}$$

At 1823 K (1550 °C), $\Delta G^\circ = 45954\,\text{J}\cdot\text{mol}^{-1}$.
Therefore,

$$\log K = -1.354$$

Hence,

$$[\text{mass\%N}] = 0.0280\,\text{mass\%}$$

At 1873 K (1600 °C), $\Delta G^\circ = 48343\ \text{J} \cdot \text{mol}^{-1}$.
Therefore,

$$\log K = -1.348$$

Hence,

$$[\text{mass\%N}] = 0.0284\,\text{mass\%}$$

At 1973 K (1700 °C), $\Delta G^\circ = 50732\ \text{J} \cdot \text{mol}^{-1}$.
Therefore,

$$\log K = -1.343$$

Hence,

$$[\text{mass\%N}] = 0.0287\,\text{mass\%}$$

Example 10.12
For an Al–Si alloy,

$$\text{Si}(\ell) = \underline{\text{Si}}$$

$$\Delta G^\circ = \mu_{\underline{\text{Si}}}^{\circ(\text{H})} - \mu_{\text{Si}}^{\circ(\text{R}),(\ell)}$$

Calculate ΔG° at 1873 K (1600 °C) using the f_{Si}° value in Fig. 10.2 ($f_{\text{Si}}^\circ = 0.0011$).
Solution:
Using Eq. (4) in Example 8.1,

$$\Delta G^\circ = \mu_{\underline{\text{Si}}}^{\circ(\text{H})} - \mu_{\text{Si}}^{\circ(\text{R}),(\ell)} = RT\ \ln f_{\text{Si}}^\circ + RT\ \ln \frac{M_{\text{Fe}}}{100 M_{\text{Si}}}$$

$$M_{Si} = 28.09,\ M_{Fe} = 55.85$$

From Fig. 10.2, $f^{\circ}_{Si} = 0.0011$ at 1873 K (1600 °C).
Hence,

$$\Delta G^{\circ} = 8.314 \times 1873 \ln 0.0011 + 8.314 \times 1873 \times \ln \frac{55.85}{100 \times 28.09}$$
$$= -167000\,\mathrm{J \cdot mol^{-1}}$$

10.3 Calculating the Standard Gibbs Energy Change, ΔG°, of Reactions

As was discussed in previous sections, it is important to know the value of ΔG° in order to ascertain equilibrium positions. In this section, we will discuss how to obtain the value of ΔG°.

10.3.1 Using the Free-Energy Function and Enthalpy

The data in data books is often given in the form $(G^{\circ}_T - H^{\circ}_0)/T$, which is termed the *free-energy function*.

Here, G°_T is the standard Gibbs energy at temperature T and H°_0 is the standard enthalpy at 0 K.

Firstly, we will derive the free-energy function,

$$\frac{G^{\circ}_T - H^{\circ}_0}{T} = \int_0^T (H^{\circ}_T - H^{\circ}_0)\mathrm{d}\left(\frac{1}{T}\right) \tag{10.66}$$

where H°_T is the standard enthalpy at temperature T.

According to the Gibbs-Helmholtz equation,

$$\left[\frac{\partial (G/T)}{\partial T}\right]_P = -\frac{H}{T^2} \tag{10.67}$$

Gibbs energy, G, and enthalpy, H, are state functions. Therefore,

$$\left[\frac{\partial(\Delta G^\circ /T)}{\partial T}\right]_P = -\frac{\Delta H^\circ}{T^2} \tag{10.68}$$

where ΔG° is the standard Gibbs energy change and ΔH° is the *standard enthalpy change*.

From Eqs. (10.68) and (10.69)

$$\frac{\mathrm{d}(1/T)}{\mathrm{d}T} = -\frac{1}{T^2} \tag{10.69}$$

we obtain

$$\left[\frac{\partial(\Delta G^\circ /T)}{\partial(1/T)}\right]_P = \Delta H^\circ \tag{10.70}$$

$\Delta G_0^\circ = \Delta H_0^\circ$ when $T \to 0\,\mathrm{K}$, thus

$$\left[\frac{\partial(\Delta H_0^\circ /T)}{\partial(1/T)}\right]_P = \Delta H_0^\circ \tag{10.71}$$

By subtracting Eq. (10.71) from Eq. (10.70), we obtain

$$\left[\frac{\partial((\Delta G^\circ - \Delta H_0^\circ)/T)}{\partial(1/T)}\right]_P = \Delta H^\circ - \Delta H_0^\circ \tag{10.72}$$

By integrating this equation from 0 to T K, we obtain

$$\frac{\Delta G^\circ - \Delta H_0^\circ}{T} - \lim_{T\to 0}\left(\frac{\Delta G^\circ - \Delta H_0^\circ}{T}\right) = \int_0^T (\Delta H^\circ - \Delta H_0^\circ)\mathrm{d}\left(\frac{1}{T}\right) \tag{10.73}$$

According to the third law of thermodynamics, the second term of the left-hand side of the equation is

$$\lim_{T\to 0}\left(\frac{\Delta G^\circ - \Delta H_0^\circ}{T}\right) = -\Delta S_0^\circ = 0 \tag{10.74}$$

where ΔS_0° is the standard entropy change at 0 K.

Thus,

$$\frac{\Delta G^\circ - \Delta H_0^\circ}{T} = \int_0^T (\Delta H^\circ - \Delta H_0^\circ)\mathrm{d}\left(\frac{1}{T}\right) \tag{10.75}$$

Therefore, for each material,

Table 10.1 Thermodynamic properties of the reaction described in Eq. (10.77)

Component	$CO_2(g)$	CO(g)	$O_2(g)$
$(G_T^\circ - H_0^\circ)/T$ [$J \cdot mol^{-1} \cdot K^{-1}$]	−226.4	−204.1	−212.1
$\Delta H_{0,i}^\circ$ [$J \cdot mol^{-1}$]	−393170	−113810	0

$$\frac{G_T^\circ - H_0^\circ}{T} = \int_0^T (H_T^\circ - H_0^\circ) \mathrm{d}\left(\frac{1}{T}\right) \tag{10.76}$$

(End of the derivation of Eq. (10.66)).

The enthalpy term $(H_T^\circ - H_0^\circ)$ can be determined using heat capacity at constant pressure, C_P°. Hence, the value of the right-hand side of the equation can be calculated if C_P° and heat of transformation are known for the temperature range 0 to T. The integrated value (the right-hand side of the equation) corresponds to the free-energy function, $(G_T^\circ - H_0^\circ)/T$. Heat capacity at constant pressure, C_P°, is used to calculate not only the free-energy function but several other thermodynamic properties, including enthalpy, H, and entropy, S.

Let us take the following reaction as an example of the calculation of ΔG° using the free-energy function:

$$CO(g) + \frac{1}{2}O_2(g) = CO_2(g) \tag{10.77}$$

The values in Table 10.1 are given for $T = 1000$ K.

The standard Gibbs energy change of the above reaction at 1000 K, ΔG_{1000}°, can be calculated as follows:

$$\frac{G_T^\circ}{T} = \frac{G_T^\circ}{T} - \frac{H_0^\circ}{T} + \frac{H_0^\circ}{T} = \frac{(G_T^\circ - H_0^\circ)}{T} + \frac{H_0^\circ}{T} \tag{10.78}$$

Thus,

$$\frac{\Delta G_{1000}^\circ}{1000} = \frac{\Delta(G_{1000}^\circ - H_0^\circ)}{1000} + \frac{\Delta H_0^\circ}{1000} \tag{10.79}$$

According to Hess's law, $\Delta H_0^\circ = \Delta(\Delta H_{0,i}^\circ)$.

Here, $\Delta H_{0,i}^\circ$ is the *standard heat of formation* of component i (also known as the *standard enthalpy of formation* of component i).

$$\frac{\Delta G^\circ_{1000}}{1000} = \frac{\Delta\left(G^\circ_{1000} - H^\circ_0\right)}{1000} + \frac{\Delta\left(\Delta H^\circ_{0,i}\right)}{1000}$$
$$= (-226.4) - (-204.1) - 1/2(-212.1)$$
$$+ [-393170 - (-113810) - (0)]/1000 \quad (10.80)$$

Thus,

$$\Delta G^\circ_{1000} = -195.63\,\text{kJ}\cdot\text{mol}^{-1} \quad (10.81)$$

The free-energy function is sometimes given in the form $\left(G^\circ_T - H^\circ_{298}\right)/T$. In this case, the equation becomes

$$\frac{\Delta G^\circ_T}{T} = \frac{\Delta\left(G^\circ_T - H^\circ_{298}\right)}{T} + \frac{\Delta\left(\Delta H^\circ_{298,i}\right)}{T} \quad (10.82)$$

In Example 10.2, ΔG° of the above reaction is given as a linear function of T, and $\Delta G^\circ_{1000} = -195.60\,\text{kJ}\cdot\text{mol}^{-1}$. Comparing this value of ΔG°_{1000} with that calculated using the free-energy function ($-195.63\,\text{kJ}\cdot\text{mol}^{-1}$) we find that the difference is small, and can be regarded as being within the bounds of measurement error for heat capacity values used to obtain the free-energy function values. Therefore, it is convenient to treat ΔG°_{1000} as a function of temperature.

10.3.2 Using Standard Gibbs Energy of Formation, $\Delta_f G^\circ_i$

The standard Gibbs energy change, ΔG°, can be described in terms of the standard Gibbs energies of formation of the various compounds in the reaction mixture.

The reaction that forms compound X_i from its elements can be described as follows:

$$X_i - \sum_k \nu_k^{\prime(i)} \varepsilon_k = 0 \quad (10.83)$$

where ν'_k is the number of moles of element k and ε_k is 1 mol of element k.

The standard Gibbs energy of formation, $\Delta_\text{f} G^\circ_i$, of coumpound X_i can be written as

$$\Delta_\text{f} G^\circ_i = \mu^\circ_i - \sum_k \nu_k^{\prime(i)} \mu^\circ_k \quad (10.84)$$

where μ°_i is the standard chemical potential of component i and μ°_k is the chemical potential of pure element ε_k in its stable state of aggregation at temperature T and pressure P.

By substituting the above equation into the following equation (which is the same as Eq. (10.4)),

$$\Delta G^\circ = \sum_{i=1}^{r} \nu_i \mu_i^\circ \tag{10.85}$$

we obtain

$$\Delta G^\circ = \sum_{i=1}^{r} \nu_i \Delta_f G_i^\circ + \sum_k \left(\sum_{i=1}^{r} \nu_k^{\prime(i)} \nu_i \right) \mu_k^\circ \tag{10.86}$$

For the difference between ν_i and ν_i', see Sects. 7.4 and 10.2.

Mass can be neither created nor destroyed in a system, except for as a result of a nuclear reaction. Therefore,

$$\sum_{i=1}^{r} \nu_k^{\prime(i)} \nu_i = 0 \tag{10.87}$$

Hence,

$$\Delta G^\circ = \sum_{i=1}^{r} \nu_i \Delta_f G_i^\circ \tag{10.88}$$

Thus, the standard Gibbs energy change, ΔG°, of the reaction can be calculated using the values for the standard Gibbs energy of formation, $\Delta_f G_i^\circ$, of each compound involved in the reaction. Similarly, the standard Gibbs energy change of a desired reaction can be calculated by combining the standard Gibbs energy change values of various reactions.

Gibbs energy, G, is, along with H, a state function. Therefore, the standard Gibbs energy of formation of a given reaction can be calculated using the standard Gibbs energy of formation of the compounds in the same manner as Hess's law.

Let us take the following reaction as an example of calculating ΔG° using the standard Gibbs energy of formation, $\Delta_f G_i^\circ$:

$$Al_2O_3(s) + 3C(s) = 2Al(\ell) + 3CO(g) \tag{10.89}$$

Since the possible combinations of components involved in reactions are near-infinite, it is impossible to list all of the data relating to standard Gibbs energy. It is also unlikely that the ΔG° value of the above reaction would be found in a data book. However, the values of the standard Gibbs energy of formation, $\Delta_f G_i^\circ$, of the reactions described below (Eqs. (10.90) and (10.91)) can be found in a data book.

$$2Al(\ell) + \frac{3}{2}O_2(g) = Al_2O_3(s) \tag{10.90}$$

$$C(s) + \frac{1}{2}O_2(g) = CO(g) \tag{10.91}$$

The $\Delta_f G_i^\circ$ values of reactions described in Eqs. (10.89)–(10.91) are:
Equation (10.89): $\Delta G^\circ_{(10.89)} = 2\mu_{Al}^{\circ,(\ell)} + 3\mu_{CO}^{\circ,(g)} - \left(\mu_{Al_2O_3}^{\circ,(s)} + 3\mu_C^{\circ,(s)}\right)$
Equation (10.90): $\Delta_f G^\circ_{Al_2O_3} = \mu_{Al_2O_3}^{\circ,(s)} - \left(2\mu_{Al}^{\circ,(\ell)} + 2/3\mu_{O_2}^{\circ,(g)}\right)$
Equation (10.91): $\Delta_f G^\circ_{f,CO} = \mu_{CO}^{\circ,(g)} - \left(\mu_C^{\circ,(s)} + 1/2\mu_{O_2}^{\circ,(g)}\right)$
As G is a state function and so is decided by the initial and final states of a system, it does not depend on the path of the change. Therefore, $\Delta G^\circ_{(10.89)}$ can be obtained from $\Delta_f G^\circ_{Al_2O_3}$ and $\Delta_f G^\circ_{CO}$ as follows:
Reaction described in Eq. (10.89) = 3 × Reaction described in Eq. (10.91)—Reaction described in Eq. (10.90)

$$\Delta G^\circ_{(10.89)} = 3 \times \Delta_f G^\circ_{CO} - \Delta_f G^\circ_{Al_2O_3} \tag{10.92}$$

Similarly, ΔG° of a reaction can be obtained by combining the $\Delta_f G_i^\circ$ of each compound involved in the reaction. By combining the formation reaction of each compound and describing the desired chemical reaction, ΔG° can be obtained as shown in the above example.

Example 10.13

Ascertain the equilibrium position of the following reaction at 1773 K (1500 °C) and 2073 K (1800 °C):

$$MgO(s) + C(s) = Mg(g) + CO(g)$$

ΔG° of the above reaction can be described as follows:

$$\Delta G^\circ = \mu_{Mg}^{\circ,(g)} + \mu_{CO}^{\circ,(g)} - \left(\mu_{MgO}^{\circ,(s)} + \mu_C^{\circ,(s)}\right)$$

$\Delta_f G^\circ_{MgO}$ of the reaction

$$\frac{1}{2}O_2(g) + Mg(g) = MgO(s) \tag{1}$$

is

$$\Delta_f G^\circ_{MgO} = -759810 - 30.84T\ \log T + 316.7T\ \mathrm{J \cdot mol^{-1}}$$

$\Delta_f G^\circ_{CO}$ of the reaction

$$C(s) + \frac{1}{2}O_2(g) = CO(g) \quad (2)$$

is

$$\Delta_f G^\circ_{CO} = -111710 - 87.65T\ \mathrm{J \cdot mol^{-1}}$$

Therefore,

$$\Delta G^\circ = \Delta_f G^\circ_{CO} - \Delta_f G^\circ_{MgO}$$
$$= 648100 + 30.84T\ \log T - 404.4T\ \mathrm{J \cdot mol^{-1}}$$

$\Delta G^\circ = 108750\,\mathrm{J \cdot mol^{-1}}$ at 1773 K (1500 °C).
In an equilibrium state ($\Delta G = 0$), $\Delta G^\circ = -RT\ \ln K$. Thus,

$$K = \exp\left(\frac{-\Delta G^\circ}{RT}\right) = \exp\left[\frac{-108750}{8.314 \times (1500 + 273)}\right] = 6.25 \times 10^{-4}$$

and

$$K = \frac{P_{Mg} P_{CO}}{a_{MgO} a_C} \approx P_{Mg} P_{CO}$$

Hence,

$$P_{Mg} P_{CO} = 6.25 \times 10^{-4}$$

Assuming $P_{Mg} = P_{CO}$, $P_{Mg} = P_{CO} = 2.5 \times 10^{-2}$bar $= 20$ mmHg.
Similarly, $P_{Mg} = P_{CO} = 0.53$ bar at 1800 °C.
At 2073 K (1800 °C), $P_{Mg} + P_{CO} > 1$ bar. Thus, the reduction reaction of MgO by carbon occurs spontaneously at atmospheric pressure (1 bar).

Example 10.14

(1) Calculate ΔG° of the following reaction using the $\Delta_f G^\circ_i$ values in Table 13.1:

$$Al_2O_3(s) + 3C(s) = 2Al(\ell) + 3CO(g)$$

(2) Calculate the P_{CO} at which Al_2O_3 begins to be reduced at 1773 K (1500 °C) and 1873 K (1600 °C). Assume that the only gas phase present is CO gas.

Solutions:
(1)
$\Delta_f G^\circ_{Al_2O_3}$ of the reaction

$$2Al(\ell) + \frac{3}{2}O_2(g) = Al_2O_3(s)$$

is

$$\Delta_f G^\circ_{Al_2O_3} = -1682900 + 324.45T \text{ J} \cdot \text{mol}^{-1}$$

$\Delta_f G^\circ_{CO}$ of the reaction

$$C(s) + \frac{1}{2}O_2(g) = CO(g)$$

is

$$\Delta_f G^\circ_{CO} = -117800 - 84.185T \text{ J} \cdot \text{mol}^{-1}$$

Hence,

$$\Delta G^\circ = 3\Delta_f G^\circ_{CO} - \Delta_f G^\circ_{Al_2O_3} = 1329500 - 577.01T \text{ J} \cdot \text{mol}^{-1}$$

(2)
The equilibrium constant, K, is

$$K = \frac{a^2_{Al} P^3_{CO}}{a_{Al_2O_3} a^3_C} = P^3_{CO}$$

$$\Delta G^\circ = -RT \ln K = -RT \ln P^3_{CO} = 1329500 - 577.01T \text{ J} \cdot \text{mol}^{-1}$$

Therefore,

$$\ln P_{CO} = -\frac{53304}{T} - 23.134$$

Hence, at 1773 K (1500 °C), Al_2O_3 is reduced when P_{CO} is lower than 7.88×10^{-24} bar.
At 1873 K (1600 °C), Al_2O_3 is reduced when P_{CO} is lower than 3.92×10^{-23} bar.

The value of ΔG° of a reaction can be calculated using not only the combination of $\Delta_f G^\circ_i$ values, but through the combination of the reactions for which ΔG° values are known.

Example 10.15

Calculate the standard Gibbs energy of the following reaction:

$$Si_3N_4(s) = 3\underline{Si} + 4\underline{N} \quad (1)$$

ΔG° of this reaction is

$$\Delta G_1^\circ = 3\mu_{\underline{Si}}^{\circ(H)} + 4\mu_{\underline{N}}^{\circ(H)} - \mu_{Si_3N_4}^{\circ,(s)}$$

By combining the following three reactions,

$$Si_3N_4(s) = 3Si(\ell) + 2N_2(g) \quad (2)$$

$$\Delta G_2^\circ = 3\mu_{Si}^{\circ,(\ell)} + 2\mu_{N_2}^{\circ,(g)} - \mu_{Si_3N_4}^{\circ,(s)}$$

$$\frac{1}{2}N_2(g) = \underline{N} \quad (3)$$

$$\Delta G_3^\circ = \mu_{\underline{N}}^{\circ(H)} - \frac{1}{2}\mu_{N_2}^{\circ,(g)}$$

$$Si(\ell) = \underline{Si} \quad (4)$$

$$\Delta G_4^\circ = \mu_{\underline{Si}}^{\circ(H)} - \mu_{Si}^{\circ,(\ell)}$$

ΔG_1° can be expressed by

$$\Delta G_1^\circ = 3\Delta G_4^\circ + 4\Delta G_3^\circ + \Delta G_2^\circ$$

According to a data book,

$$\Delta G_2^\circ = 874460 - 405.0T \text{ J} \cdot \text{mol}^{-1}$$

$$\Delta G_3^\circ = 3598 + 23.89T \text{ J} \cdot \text{mol}^{-1}$$

$$\Delta G_4^\circ = -131500 - 17.24T \text{ J} \cdot \text{mol}^{-1}$$

Hence,

$$\Delta G_1^\circ = 494650 - 361.2T \text{ J} \cdot \text{mol}^{-1}$$

Example 10.16

Calculate the $a_{Si}^{(H)}$ (Henrian (1 mass% i) standard state) value required to stop the dissolution of Si_3N_4 into molten iron at 1873 K (1600 °C).

Solution:

The reaction can be described as follows:

$$Si_3N_4(s) = 3\underline{Si} + 2N_2(g) \qquad (1)$$

ΔG° of this reaction is

$$\Delta G_1^\circ = 3\mu_{\underline{Si}}^{\circ(H)} + 2\mu_{N_2}^{\circ,(g)} - \mu_{Si_3N_4}^{\circ,(s)}$$

As shown in Example 10.15, according to a data book,

$$\Delta G_4^\circ = \mu_{\underline{Si}}^{\circ(H)} - \mu_{Si}^{\circ,(\ell)} = -131500 - 17.24T \text{ J} \cdot \text{mol}^{-1}$$

$$\Delta G_2^\circ = 3\mu_{Si}^{\circ,(\ell)} + 2\mu_{N_2}^{\circ,(g)} - \mu_{Si_3N_4}^{\circ,(s)} = 874460 - 405.0T \text{ J} \cdot \text{mol}^{-1}$$

Therefore,

$$\Delta G_1^\circ = 3\Delta G_4^\circ + \Delta G_2^\circ = 479960 - 456.72T \text{ J} \cdot \text{mol}^{-1}$$

Hence,

$$\begin{aligned}\Delta G_1 &= \Delta G_1^\circ + RT \ln K \\ &= 479960 - 456.72T + RT \ln \frac{a_{Si}^{(H)3} P_{N_2}^2}{a_{Si_3N_4}} \\ &= 479960 - 456.72T + RT \ln a_{Si}^{(H)3}\end{aligned}$$

Dissolution stops when $\Delta G_1 = 0$, i.e.

$$\ln a_{Si}^{(H)} = -\frac{19243}{T} + 18.311$$

At 1873 K (1600 °C),

$$\ln a_{\mathrm{Si}}^{(\mathrm{H})} = 8.037$$

$$a_{\mathrm{Si}}^{(\mathrm{H})} = 3093$$

This value for $a_{\mathrm{Si}}^{(\mathrm{H})}$ is quite high. However, if we convert it to $a_{\mathrm{Si}}^{(\mathrm{R})}$ using Eq. (8.72), $a_{\mathrm{Si}}^{(\mathrm{R})} = 0.068$. According to Fig. 10.2, the corresponding x_{Si} value is $x_{\mathrm{Si}} = 0.36$.

10.3.3 Using Heat Capacity

The change in heat capacity at constant pressure, ΔC_P, caused by the following reaction

$$\nu_1' \mathrm{X}_1 + \nu_2' \mathrm{X}_2 = \nu_3' \mathrm{X}_3 + \nu_4' \mathrm{X}_4 \tag{10.93}$$

where $\mathrm{X}_i\,(i = 1, 2, 3, 4)$ is 1 mole of each component and $\nu_i'\,(i = 1, 2, 3, 4)$ is the number of moles of each component, is expressed by

$$\Delta C_P = \nu_3' \overline{C}_{P,3} + \nu_4' \overline{C}_{P,4} - \left(\nu_1' \overline{C}_{P,1} + \nu_2' \overline{C}_{P,2}\right) = \Delta \mathrm{a} + \Delta \mathrm{b} T + \Delta \mathrm{c} T^{-2} \tag{10.94}$$

where $\overline{C}_{P,i}$ is the molar heat capacity (heat capacity per mole) of component i.

ΔG° of the above reaction can be expressed as follows:

$$\Delta G^\circ = \Delta H_0^{\circ\prime} - \Delta \mathrm{a} T \ln T - \frac{1}{2} \Delta \mathrm{b} T^2 - \frac{1}{2} \Delta \mathrm{c} T^{-1} + \left(\Delta \mathrm{a} - \Delta S_0^{\circ\prime}\right) T \tag{10.95}$$

where $\Delta H_0^{\circ\prime}$ and $\Delta S_0^{\circ\prime}$ are integral constants, and Δa, Δb, and Δc are coefficients.

Equation (10.95) can be derived as follows:

The Gibbs energy of States 1 and 2 can be expressed as follows:

$$G_1 = H_1 - T S_1 \tag{10.96}$$

$$G_2 = H_2 - T S_2 \tag{10.97}$$

where G is the Gibbs energy, H is the enthalpy, T is the temperature, and S is the entropy.

When the system changes from State 1 to State 2 as a result of a reaction, the Gibbs energy change, ΔG, is

$$\Delta G = G_2 - G_1 = (H_2 - H_1) - T(S_2 - S_1) = \Delta H - T\Delta S \tag{10.98}$$

The change in standard state is

$$\Delta G^\circ = \Delta H^\circ - T\Delta S^\circ \tag{10.99}$$

where ΔG° is the standard Gibbs energy change, ΔH° is the standard enthalpy change, and ΔS° is the standard entropy change. As can be seen from the above equation, ΔG° can be calculated using ΔH° and ΔS°.

According to Kirchhoff's law (Eq. (4.21)), the standard enthalpy change at temperature T, ΔH_T°, as a result of a reaction can be described in terms of *heat capacity change* in a standard state ($P = 1$ bar), ΔC_P°, as follows:

$$\Delta H_T^\circ = \int_{T^\circ}^{T} \Delta C_P^\circ \mathrm{d}T + \Delta H_{T^\circ}^\circ \tag{10.100}$$

Based on Eqs. (6.14) and (6.15), the standard entropy change at temperature T,ΔS_T°, can be described as a function of ΔC_P° as follows:

$$\Delta S_T^\circ = \int_{T^\circ}^{T} \frac{\Delta C_P^\circ}{T} \mathrm{d}T + \Delta S_{T^\circ}^\circ \tag{10.101}$$

Hence, the standard enthalpy change and the standard entropy change at temperature T can be calculated if the values at T° ($\Delta H_{T^\circ}^\circ$ and $\Delta S_{T^\circ}^\circ$) are known.

When C_P is expressed by

$$C_P = \mathrm{a} + \mathrm{b}T + \mathrm{c}T^{-2} \tag{10.102}$$

ΔC_P° can be expressed by

$$\Delta C_P^\circ = \Delta\mathrm{a} + \Delta\mathrm{b}T + \Delta\mathrm{c}T^{-2} \tag{10.103}$$

By integrating ΔC_P° with respect to T, we obtain

$$\Delta H^\circ = \Delta H_0^{\circ\prime} + \Delta\mathrm{a}T + \frac{1}{2}\Delta\mathrm{b}T^2 - \Delta\mathrm{c}T^{-1} \tag{10.104}$$

where $\Delta H_0^{\circ\prime}$ is an integral constant.

By integrating $\frac{\Delta C_P^\circ}{T}$ with respect to T, we obtain

$$\Delta S^{\circ} = \Delta S_0^{\circ\prime} + \Delta\mathrm{a} \ln T + \Delta\mathrm{b}T - \frac{1}{2}\Delta\mathrm{c}T^{-2} \tag{10.105}$$

where $\Delta S_0^{\circ\prime}$ is an integral constant.

By substituting Eqs. (10.104) and (10.105) into Eq. (10.99), the following equation is obtained:

$$\Delta G^\circ = \Delta H_0^{\circ\prime} - \Delta \mathrm{a} T \ln T - \frac{1}{2}\Delta \mathrm{b} T^2 - \frac{1}{2}\Delta \mathrm{c} T^{-1} + \left(\Delta \mathrm{a} - \Delta S_0^{\circ\prime}\right)T \quad (10.106)$$

(End of the derivation of Eq. (10.95))

The relationship between ΔG° and heat capacity can be described in another form:

$$\Delta G^\circ(T) = \Delta H_0^\circ - T\int_0^T \frac{\mathrm{d}T}{T^2}\int_0^T \Delta C_P^\circ \mathrm{d}T \quad (10.107)$$

This equation can be derived as follows:

Firstly,

$$\left[\frac{\partial(\Delta G^\circ/T)}{\partial T}\right]_P = \frac{1}{T}\left(\frac{\partial \Delta G^\circ}{\partial T}\right)_P - \frac{1}{T^2}\Delta G^\circ = -\frac{\Delta S^\circ}{T} - \frac{\Delta G^\circ}{T^2} = \frac{-\Delta H^\circ}{T^2} \quad (10.108)$$

Equation (10.108) is a form of the Gibbs-Helmholtz equation. By integrating it from T_0 to T, we obtain

$$\frac{\Delta G^\circ(T)}{T} - \frac{\Delta G^\circ(T_0)}{T_0} = -\int_{T_0}^{T} \frac{\Delta H^\circ}{T^2}\mathrm{d}T \quad (10.109)$$

We now recall Kirchhoff's law, which relates the heat capacity change to the temperature coefficient of the heat of reaction:

$$\Delta C_P = \left(\frac{\partial \Delta H}{\partial T}\right)_P \quad (10.110)$$

where ΔC_P is the heat capacity change, ΔH is the enthalpy change, and T is the temperature.

By integrating Eq. (10.110) from T_0 to T, we obtain

$$\int_{T_0}^{T} \Delta C_P \mathrm{d}T = \Delta H(T) - \Delta H(T_0) \quad (10.111)$$

Hence,

$$\Delta H(T) = \Delta H(T_0) + \int_{T_0}^{T} \Delta C_P \mathrm{d}T \tag{10.112}$$

In a standard state,

$$\Delta H^\circ = \Delta H^\circ(T_0) + \int_{T_0}^{T} \Delta C_P^\circ \mathrm{d}T \tag{10.113}$$

Therefore,

$$\frac{\Delta G^\circ(T)}{T} - \frac{\Delta G^\circ(T_0)}{T_0} = \Delta H^\circ(T_0)\left(\frac{1}{T} - \frac{1}{T_0}\right) - \int_{T_0}^{T} \frac{\mathrm{d}T}{T^2} \int_{T_0}^{T} \Delta C_P^\circ \mathrm{d}T \tag{10.114}$$

By substituting

$$\Delta G^\circ(T_0) = \Delta H^\circ(T_0) - T_0 \Delta S^\circ(T_0) \tag{10.115}$$

into the above equation, we obtain

$$\frac{\Delta G^\circ(T)}{T} = \frac{\Delta H^\circ(T_0) - T\Delta S^\circ(T_0)}{T} - \int_{T_0}^{T} \frac{\mathrm{d}T}{T^2} \int_{T_0}^{T} \Delta C_P^\circ \mathrm{d}T \tag{10.116}$$

According to the third law of thermodynamics, $\Delta S^\circ(T_0) = 0$ at $T_0 = 0\,\mathrm{K}$. Hence,

$$\Delta G^\circ(T) = \Delta H_0^\circ - T \int_0^T \frac{\mathrm{d}T}{T^2} \int_0^T \Delta C_P^\circ \mathrm{d}T \tag{10.117}$$

(End of the derivation of Eq. (10.107))

Therefore, if the values for ΔH_0° and ΔC_P° are given as a function of temperature, $\Delta G^\circ(T)$ can be calculated at temperature T. In this sense, Eq. (10.107) differs from Eq. (10.95) as the latter includes two integral constants, $\Delta S_0^{\circ\prime}$ and $\Delta H_0^{\circ\prime}$.

Example 10.17

Calculate the dissociation pressure of oxygen for Al_2O_3(s) at 700 K. The reaction is $2Al(s) + \frac{3}{2}O_2(g) = Al_2O_3(s)$. Use the following data for the calculation. The units of $\Delta H^\circ, \overline{C}_{P,i}^\circ, \overline{S}_i^\circ$, and ΔG° are $\mathrm{J \cdot mol^{-1}}$, $\mathrm{J \cdot mol^{-1} \cdot K^{-1}}$, $\mathrm{J \cdot mol^{-1} \cdot K^{-1}}$, and $\mathrm{J \cdot mol^{-1}}$, respectively.

$$\Delta H^\circ_{298} = -1675300$$

$$\overline{C}^\circ_{P,\mathrm{Al_2O_3}} = 114.8 + 12.80 \times 10^{-3}T - 35.44 \times 10^5 T^{-2}\ (273 - 2000\ \mathrm{K})$$

$$\overline{C}^\circ_{P,\mathrm{Al}} = 20.67 + 12.38 \times 10^{-3}T\,(273 - 923\ \mathrm{K})$$

$$\overline{C}^\circ_{P,\mathrm{O_2}} = 29.96 + 4.184 \times 10^{-3}T - 1.67 \times 10^5 T^{-2}(273 - 2000\ \mathrm{K})$$

$$\overline{S}^\circ_{298,\mathrm{Al_2O_3}} = 51.21$$
$$\overline{S}^\circ_{298,\mathrm{Al}} = 28.33$$
$$\overline{S}^\circ_{298,\mathrm{O_2}} = 205.0$$

Solution:
Using the above data, ΔC°_P can be described as follows:

$$\begin{aligned}\Delta C^\circ_P &= \overline{C}^\circ_{P,\mathrm{Al_2O_3}} - \left(\frac{3}{2}\overline{C}^\circ_{P,\mathrm{O_2}} + 2\overline{C}^\circ_{P,\mathrm{Al}}\right)\\ &= 28.52 - 18.24 \times 10^{-3}T - 32.93 \times 10^5 T^{-2}\end{aligned}$$

By integrating it with respect to T, we obtain

$$\Delta H^\circ = \Delta H^{\circ\prime}_0 + 28.52T - \frac{18.24}{2} \times 10^{-3}T^2 + 32.93 \times 10^5 T^{-1}$$

$\Delta H^\circ_{298} = -1675300$ at $T = 298.15$ K. Thus, $\Delta H^{\circ\prime}_0 = -1694000$
By integrating $\frac{\Delta C^\circ_P}{T}$ with respect to T, we obtain

$$\Delta S^\circ = \Delta S^{\circ\prime}_0 + 28.52 \ln T - 18.24 \times 10^{-3}T + \frac{32.93}{2} \times 10^5 T^{-2}$$

$\Delta S^\circ_{298} = -313.0$ at $T = 298.15$ K. Thus $\Delta S^{\circ\prime}_0 = -488.6$
Hence,

$$\begin{aligned}\Delta G^\circ = \Delta H^\circ - T\Delta S^\circ &= -1694100 - 28.52T\ \ln T\\ &+ 9.12 \times 10^{-3}T^2 + 16.48 \times 10^5 T^{-1} + 516.93T\\ &= -1456200\ \text{at}\ 700\ \mathrm{K}.\end{aligned}$$

$$K = \frac{a_{Al_2O_3}}{a_{Al}^2 P_{O_2}^{3/2}} \approx \frac{1}{P_{O_2}^{3/2}}$$

$$\Delta G^\circ = -RT\ \ln K$$

Thus,

$$\Delta G^\circ = -8.314 \times 700 \ln\left(\frac{1}{P_{O_2}^{3/2}}\right) = -1456200$$

Therefore,

$$P_{O_2} = 3.6 \times 10^{-73}\ \text{bar}$$

The equation for values of ΔG° used in the above example is relatively complex, and the application of the thermochemical properties used to derive ΔG° involves errors to a certain extent. Thus, the use of a simpler equation to approximate ΔG° is, in practice, often both possible and more convenient.

Approximation (1): $\Delta C_P^\circ = 0$

$\Delta a = \Delta b = \Delta c = 0$ in Eq. (10.95).
Hence,

$$\Delta G^\circ = \Delta H_0^{\circ\prime} - \Delta S_0^{\circ\prime} T \tag{10.118}$$

Approximation (2): $\Delta C_P^\circ = \Delta a$ (constant)

$$\Delta b = \Delta c = 0 \tag{10.119}$$

Thus,

$$\Delta G^\circ = \Delta H_0^{\circ\prime} - \Delta a T\ \ln T + \left(\Delta a - \Delta S_0^{\circ\prime}\right) T \tag{10.120}$$

Approximation (3): A linear fitting of Eq. (10.95)

A linear approximation can be applied to Eq. (10.95) in the temperature range within which ΔC_P° can be regarded as zero:

$$\Delta G_T^\circ = \Delta H^\circ - \Delta S^\circ T \tag{10.121}$$

where ΔH° and ΔS° are the average standard enthalpy change and the average standard entropy change within the temperature range, respectively. The equation is simple, but sufficient to give values that are close to those calculated using Eq. (10.121),

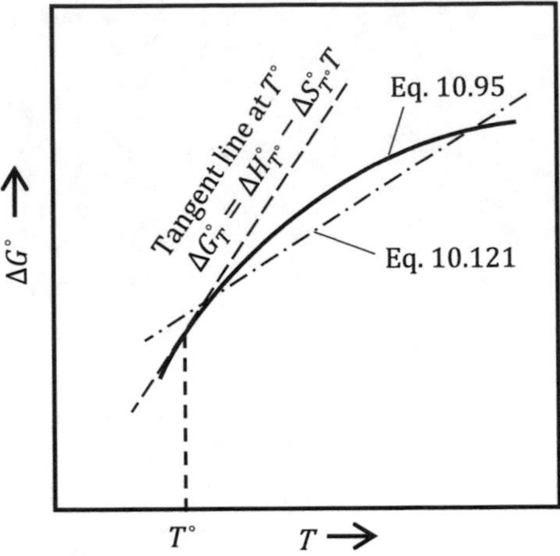

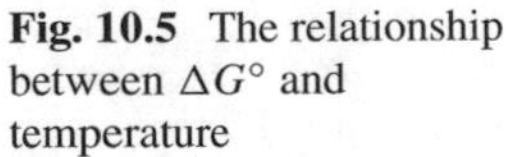
Fig. 10.5 The relationship between ΔG° and temperature

which is derived using complex ΔG° equations. A schematic illustration of the relationship between ΔG° and temperature is shown in Fig. 10.5.

10.4 The Relationship Between Equilibrium Constant, Temperature, and Pressure

In this section, we will explore the relationship between equilibrium constant, temperature, and pressure. The equilibrium constant, K, is defined by Eqs. (10.7) and (10.8), and can be expressed by the following equation (cf. Eq. (10.6)):

$$K = \exp\left(\frac{-\Delta G^\circ}{RT}\right) \tag{10.122}$$

where ΔG° is the standard Gibbs energy change, R is the gas constant, and T is the temperature.

This equation is another definition of the equilibrium constant. In the case of a gas, $\Delta G^\circ = \sum_{i=1}^{r} \nu_i \mu_i^{\circ(\mathrm{g})}$ and $\mu_i^{\circ(\mathrm{g})}$ is a function only of temperature. Therefore, based on Eq. (10.122), the equilibrium constant is also a function only of temperature. In the case of a liquid, if $\mu_i^\circ(T, P)$ is taken as a standard state (using Eqs. (10.4) and (10.122)), the equilibrium constant is a function of temperature and pressure; if $\mu_{\underline{i}}^{\circ(\mathrm{H})}$ is taken as a standard state, the equilibrium constant depends on the solvents as well as temperature and pressure.

10.4.1 Temperature Dependence of the Equilibrium Constant

We will now consider the temperature dependence of the equilibrium constant at constant pressure. Based on Eqs. (10.4) and (10.6),

$$\ln K = \frac{-\Delta G^\circ}{RT} = -\sum_{i=1}^{r} \frac{\nu_i \mu_i^\circ}{RT} \tag{10.123}$$

where ν_i is the stoichiometric number of the reaction (which is negative for reactants and positive for products) and μ_i° is the standard chemical potential of component i.

By differentiating the above equation with respect to T,

$$\left(\frac{\partial \ln K}{\partial T}\right)_P = -\sum_{i=1}^{r} \frac{\nu_i}{R}\left[\frac{\partial(\mu_i^\circ/T)}{\partial T}\right]_P \tag{10.124}$$

The term $\left[\frac{\partial(\mu_i^\circ/T)}{\partial T}\right]_P$ on the right-hand side of the above equation is

$$\left[\frac{\partial(\mu_i^\circ/T)}{\partial T}\right]_P = \frac{1}{T}\left(\frac{\partial \mu_i^\circ}{\partial T}\right)_P - \frac{\mu_i^\circ}{T^2} = -\frac{\overline{S}_i^\circ}{T} - \frac{\mu_i^\circ}{T^2} = \frac{-\overline{H}_i^\circ}{T^2} \tag{10.125}$$

where $\overline{S}_i^\circ$ is the standard partial molar entropy and $\overline{H}_i^\circ$ is the standard partial molar enthalpy of component i at temperature T and pressure P. By substituting Eq. (10.125) into Eq. (10.124), we obtain

$$\left(\frac{\partial \ln K}{\partial T}\right)_P = \sum_{i=1}^{r} \frac{\nu_i \overline{H}_i^\circ}{RT^2} = \frac{\Delta H^\circ}{RT^2} \tag{10.126}$$

Equation (10.126) is called the *van't Hoff equation*. ΔH° is the standard enthalpy change of the reaction, which is expressed by

$$\Delta H^\circ = \sum_{i=1}^{r} \nu_i \overline{H}_i^\circ \tag{10.127}$$

In the case of liquid/solid reactions, if Convention (a) is chosen as the standard state (see Sect. 8.2.3.2), ΔH° is the heat absorbed in the forward reaction when all of the substances involved in the reaction are in their pure states. If Convention (b) is chosen as the standard state, ΔH° is the heat absorbed in the forward reaction when the solvent is pure and all of the other substances involved in the reaction are in a hypothetical infinitely dilute state (i.e. one that satisfies $\mu_i = \mu_{\underline{i}}^{\circ(\mathrm{H})} + RT\ \ln[\mathrm{mass\%}\ i]$).

In the case of a gaseous reaction, $\overline{H}_i \approx \overline{H}_i^\circ$ as the gas can be regarded as an ideal gas up to around 1 bar, and ΔH° is nearly equal to the heat of reaction at an arbitrary concentration of gas mixture.

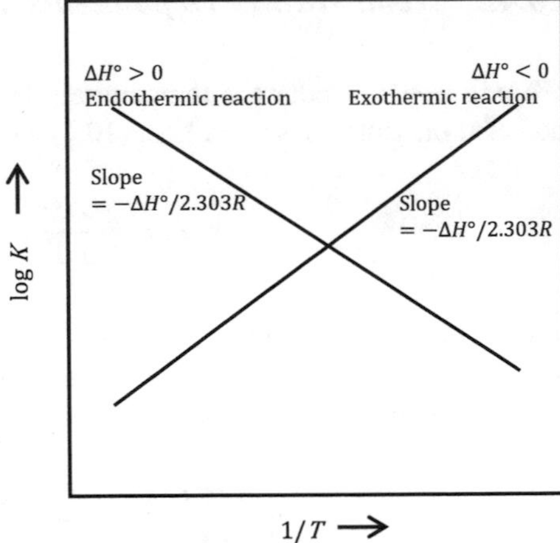

Fig. 10.6 Temperature dependence of equilibrium constant for endothermic and exothermic reactions

As is discussed in Example 4.2, the change in ΔH° in relation to temperature is small.

Therefore, by integrating Eq. (10.126) within the temperature range at which ΔH° is considered to be virtually constant, we obtain

$$\ln K = \frac{-\Delta H^\circ}{RT} + \mathrm{I} \tag{10.128}$$

where I is an integral constant.

Between the upper and lower limits of the temperature range,

$$\ln \frac{K(T_1)}{K(T_2)} = -\left(\frac{\Delta H^\circ}{R}\right)\left(\frac{1}{T_1} - \frac{1}{T_2}\right) \tag{10.129}$$

Therefore, by plotting log K against $1/T$, the slope $-\Delta H^\circ /2.303R$ is obtained, as shown in Fig. 10.6. When $\Delta H^\circ > 0$, i.e. in the case of an endothermic reaction, the equilibrium constant, K, increases. with increasing temperature; when $\Delta H^\circ < 0$, i.e. in the case of an exothermic reaction, the equilibrium constant, K, increases with decreasing temperature.

10.4.2 Pressure Dependence of the Equilibrium Constant

We will now turn to the pressure dependence of the equilibrium constant at a constant temperature. The differentiation of Eq. (10.123) with respect to P at a constant temperature is

$$\left(\frac{\partial \ln K}{\partial P}\right)_T = -\sum_{i=1}^{r} \frac{\nu_i}{RT}\left(\frac{\partial \mu_i^\circ}{\partial P}\right)_T \tag{10.130}$$

where K is the equilibrium constant, P is the pressure, ν_i is the stoichiometric number of the reaction (which is negative for reactants and positive for products), R is the gas constant, T is the temperature, and μ_i° is the standard chemical potential of component i.

Using Eq. (7.32), we obtain

$$\left(\frac{\partial \mu_i^\circ}{\partial P}\right) = \overline{V}_i^\circ \tag{10.131}$$

where $\overline{V}_i^\circ$ is the standard molar volume of component i at temperature T and pressure P.

Hence,

$$\left(\frac{\partial \ln K}{\partial P}\right)_T = -\sum_{i=1}^{r} \frac{\nu_i \overline{V}_i^\circ}{RT} = -\frac{\Delta V^\circ}{RT} \tag{10.132}$$

$$\Delta V^\circ = \sum_{i=1}^{r} \nu_i \overline{V}_i^\circ \tag{10.133}$$

where ΔV° is the standard volume change.

In the case of a gas, and as is discussed above, $\mu_i^{\circ(g)}$ is a function only of temperature.[1]

Therefore,

$$\left(\frac{\partial \ln K}{\partial P}\right)_T = 0 \tag{10.134}$$

Thus, the equilibrium constant of a gaseous reaction is not dependent on pressure when an ideal gas at 1 bar is chosen as the standard state for each component in the reactant.

In general, the equilibrium constant of a gaseous reaction is expressed as $K = \prod_{i=1}^{r} f_i^{\nu_i}$, where Π (capital pi) denotes the products (i.e. $\prod_{i=1}^{r} f_i^{\nu_i} = f_1^{\nu_1} \cdot f_2^{\nu_2} \cdot f_3^{\nu_3} \cdots f_r^{\nu_r}$) and f_i is the fugacity of component i.

---

[1]However, if the chemical potential of component i in the gas mixture is expressed as

$$\mu_i = \left[\mu_i^\circ(T) + RT \ln P\right] + RT \ln\left\{x_i \exp\left[\frac{1}{RT}\int_0^P \left(\overline{V}_i - \frac{RT}{P}\right)\right] \mathrm{d}P\right\}$$

and $\mu_i^{\circ(g)}(T) + RT \ln P$ is defined as a new standard chemical potential, this is a function of temperature and pressure. Thus, the equilibrium constant, K, also depends on (total) pressure.

10.5 Ellingham Diagram

In this section, a method of obtaining partial oxygen pressure in an equilibrium state, i.e. the dissociation pressure of oxygen as a result of a reaction between e.g. Me (metal) and O,

$$\mathrm{Me(s,\ell) + O_2(g) = MeO_2(s,\ell)} \tag{10.135}$$

using a diagram is discussed.

The feasibility of the oxidation of Me and reduction of MeO_2 in a $CO - CO_2$ mixture gas and a $H_2 - H_2O$ mixture gas will be discussed in relation to a diagram. Sulphides will also be discussed in relation to this.

This method is useful for discussing the stability and reactivity of oxides, sulphides, etc., as the difference in dissociation pressure can be compared in a simple way. The relationship between the ΔG° value of the production reaction for the oxides/sulphides and T is linear within the temperature range in which no phase transformation occurs. A diagram that shows the relationship between ΔG° and T is called an *Ellingham diagram*, and is a graphical method of determining partial oxygen pressure in an equilibrium state (dissociation pressure of oxygen), $P_{O_{2,e}}$.

10.5.1 *Obtaining the Dissociation Pressure of Oxygen*

The standard Gibbs energy change, ΔG°, of the reaction

$$\mathrm{Me(s,\ell) + O_2(g) = MeO_2(s,\ell)} \tag{10.136}$$

can be expressed as

$$\Delta G^\circ = \mu^\circ_{\mathrm{MeO_2}} - \left(\mu^\circ_{\mathrm{Me}} + \mu^{\circ,(g)}_{\mathrm{O_2}}\right) \tag{10.137}$$

where s, ℓ, and g in the brackets denote solid, liquid, and gas phases, respectively. μ_i° is the standard chemical potential of component i.

When there is no mutual solubility between Me and MeO_2, ΔG° is:

$$\Delta G^\circ = -RT\,\ln\left(\frac{1}{P_{O_{2,e}}}\right) = RT\,\ln P_{O_{2,e}} \tag{10.138}$$

where R is the gas constant, T is the temperature, and $P_{O_{2,e}}$ is the partial equilibrium oxygen pressure (dissociation pressure of oxygen).

The relationship between ΔG° and T can be plotted by taking $RT\,\ln P_{O_{2,e}}$ as a vertical axis and T as a horizontal axis (Fig. 10.7) after arranging the reaction described in Eq. (10.136) such that the coefficient of O_2 is 1. As the value of partial

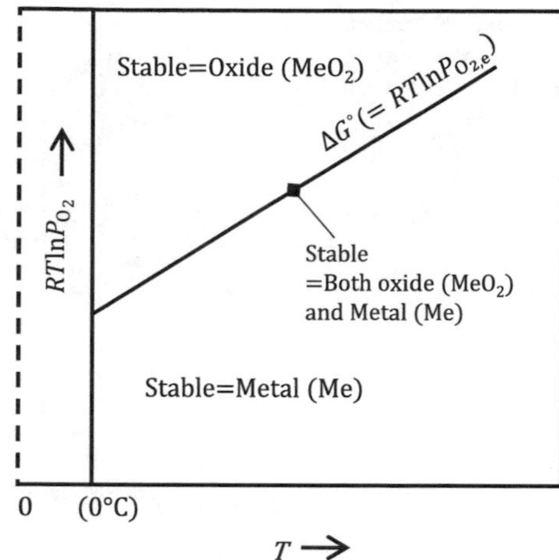

Fig. 10.7 Stable region of Me and MeO_2

oxygen pressure (P_{O_2}) on the $\Delta G^\circ - T$ line is $P_{O_{2,e}}$, Me and MeO_2 coexist at this pressure. Above this line, however, $P_{O_2} > P_{O_{2,e}}$, and thus ΔG becomes

$$\Delta G = \Delta G^\circ + RT\ \ln P_{O_2}^{-1} = RT\ \ln\left(\frac{P_{O_{2,e}}}{P_{O_2}}\right) < 0 \tag{10.139}$$

This implies that the reaction occurs spontaneously; in other words, the oxygen is consumed and the reaction occurs towards the right-hand side, such that P_{O_2} decreases and the system approaches an equilibrium state ($P_{O_{2,e}}$). If the oxygen supply is such that $P_{O_2} > P_{O_{2,e}}$ continues to be true, the reaction will continue until the Me is completely consumed. Therefore, in the P_{O_2} region above the line, Me is unstable and MeO_2 is stable.

In the region below the line, $P_{O_2} < P_{O_{2,e}}$, and the reaction occurs towards the left-hand side (decomposition of MeO_2), such that P_{O_2} increases and the system approaches an equilibrium state ($P_{O_{2,e}}$). If the condition $P_{O_2} < P_{O_{2,e}}$ is maintained by e.g. evacuating the system, Me becomes stable and MeO_2 becomes unstable.

Figure 10.8 was obtained by adding the lines for each P_{O_2} value to Fig. 10.7.

For example, the line for $P_{O_2} = 1$(bar) is $RT\ln P_{O_2} = 0$. The line for $P_{O_2} = 10^{-1}$ (bar) is $\left(R\ \ln 10^{-1}\right) \times T = -19.15T\ \mathrm{J \cdot mol^{-1}}$ i.e. the line that passes the origin and has the slope $-19.15\ \mathrm{J \cdot mol^{-1}}$. Similarly, the lines for $P_{O_2} = 10^{-2}$ (bar), 10^{-3}(bar), ··· can be drawn.

Therefore, the scale for P_{O_2} can be created by plotting the P_{O_2} values on the secondary axis.

The value of P_{O_2} on the scale corresponds to P_{O_2} at each temperature on the line between the origin and the point on the scale. For example, if we want to ascertain $P_{O_{2,e}}$ for MeO_2 at temperature T°, we first draw the vertical line that passes temperature T° and find the point that intersects with the $\Delta G^\circ - T$ line (Point A in Fig. 10.8). By drawing a horizontal line from Point A towards the $RT\ \ln P_{O_2}$ axis we obtain the

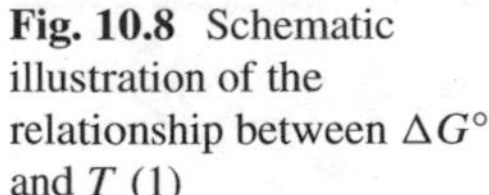

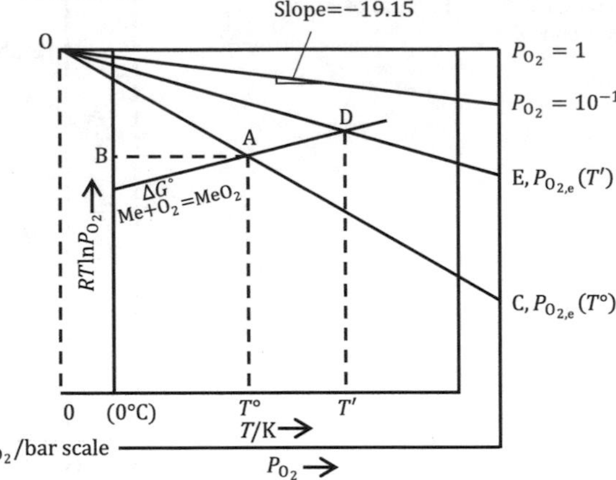

Fig. 10.8 Schematic illustration of the relationship between ΔG° and T (1)

$RT \ln P_{O_2}$ value (Point B), and so P_{O_2} can be calculated. However, as Point A lies on the line OC it is more convenient to read the P_{O_2} value (Point C) on the scale for P_{O_2} directly.

In summary, the following information can be obtained from Fig. 10.8:

(1) Dissociation pressure of oxygen for MeO_2 at temperature T'

Draw a line that passes temperature T' and find the point that intersects with the $\Delta G^\circ - T$ line (Point D). Then, draw the O–D line and extrapolate it towards the P_{O_2} scale; the intersecting point (E) is the dissociation pressure of oxygen, $P_{O_{2,e}}$.

(2) Temperature, T', at which the partial oxygen pressure is $P_{O_{2,e}}$ (dissociation pressure of oxygen)

Draw the O–E line ($P_{O_{2,e}}$) and find the point that intersects with the $\Delta G^\circ - T$ line (Point D). Then draw a vertical line towards the temperature scale. The intersecting point (T') is the temperature at which the partial oxygen pressure is $P_{O_{2,e}}$.

Ellingham diagrams for oxides are shown in Figs. 10.9 and 10.10.

Example 10.18

Determine which of the following metals do not oxidise at 1273 K (1000 °C) and $P_{O_2} = 10^{-8}$ bar using an Ellingham diagram.

Assume that the gas phase consists of only oxygen, and the oxides in the brackets are the only oxides that might be formed.

Cu (CuO), Ni (NiO), Fe (FeO), Cr (Cr_2O_3), Si (SiO_2), Al (Al_2O_3), Ca (CaO)

Solution:

In Figs. 10.9 and 10.10, the $P_{O_{2,e}}$ values of Ni, Fe, Cr, Al, and Ca at 1273 K are lower than 10^{-8} bar. Hence, these elements will oxidise. However, the $P_{O_{2,e}}$ value of Cu is higher than 10^{-8} bar, and so Cu will not oxidise.

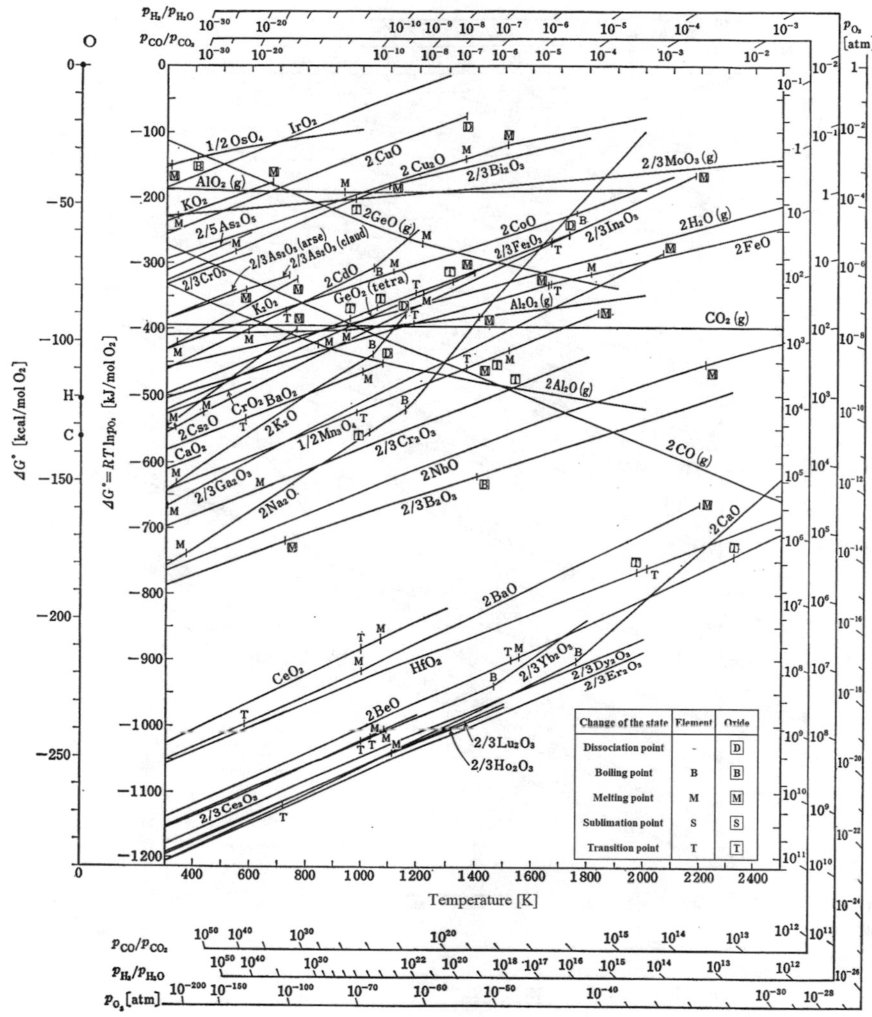

Fig. 10.9 The relationship between the standard Gibbs energy of formation of oxides and temperature (1) (*Handbook of Iron and Steel, I (Fundamentals)*, 3rd Ed., ed. by ISIJ,. Maruzen, Tokyo, 1981)

10.5.2 Equilibrium Between $CO-CO_2$ and MeO_2

The standard Gibbs energy change, ΔG°, of the reaction

$$2CO(g) + O_2(g) = 2CO_2(g) \tag{10.140}$$

is

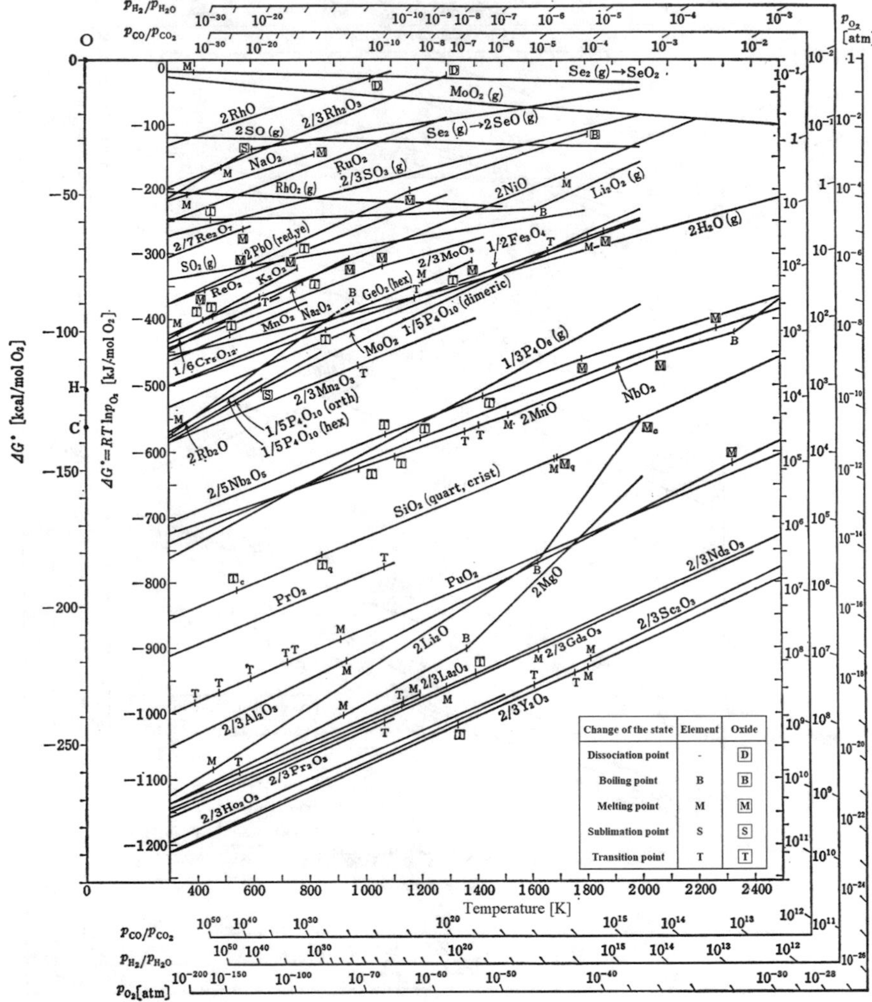

Fig. 10.10 The relationship between the standard Gibbs energy of formation of oxides and temperature (2) (*Handbook of Iron and Steel, I (Fundamentals)*, 3rd Ed., ed. by ISIJ,. Maruzen, Tokyo, 1981)

$$\Delta G^\circ = -RT\ \ln \frac{P_{CO_2}^2}{P_{CO}^2 P_{O_2}} \tag{10.141}$$

where P_i is the partial pressure of component i.

Therefore,

$$RT\ \ln P_{O_2} = \Delta G^\circ - 2RT\ \ln \frac{P_{CO}}{P_{CO_2}} \tag{10.142}$$

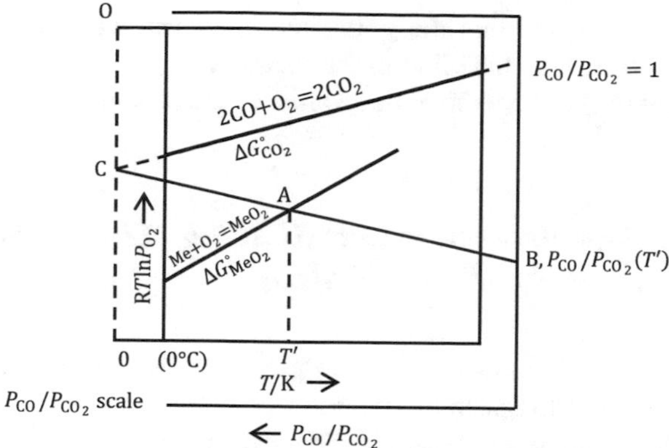

Fig. 10.11 Schematic illustration of the relationship between ΔG° and T (2)

Hence, the partial oxygen pressure of a $CO-CO_2$ system in an equilibrium state is determined by the P_{CO}/P_{CO_2} ratio, which we can investigate using an Ellingham diagram.

$RT \ln P_{O_2} = \Delta G^\circ$ when $P_{CO}/P_{CO_2} = 1$. By extrapolating this relationship to $T = 0$ K, Point C is obtained (Fig. 10.11).

From Point C, the lines for each P_{CO}/P_{CO_2} value can be drawn as follows:

$RT \ln P_{O_2} = \Delta G^\circ - 2 \times 19.15T \text{ J} \cdot \text{mol}^{-1}$ at $P_{CO}/P_{CO_2} = 10$,

$RT \ln P_{O_2} = \Delta G^\circ - 4 \times 19.15T \text{ J} \cdot \text{mol}^{-1}$ at $P_{CO}/P_{CO_2} = 10^2$, etc.

By adding another axis (as in the P_{O_2} case, see Sect. 10.5.1), the scale for the P_{CO}/P_{CO_2} can be obtained (Fig. 10.11).

Hence, the partial oxygen pressure P_{O_2} of the reaction described in Eq. (10.140) at a given P_{CO}/P_{CO_2} value can be found on the line between Point C and the P_{CO}/P_{CO_2} value on the scale. The P_{O_2} value of the reaction at temperature T can be obtained using the method described in Sect. 10.5.1.

In summary, the following information can be obtained from Fig. 10.11:

(1) The P_{CO}/P_{CO_2} value at which Me and MeO_2 coexist at temperature T'

Draw a vertical line that passes temperature T' and find the point that intersects with the $\Delta G^\circ - T$ line (Point A in Fig. 10.11). Then, draw the line A–C and extrapolate it towards the P_{CO}/P_{O_2} scale. The intersecting point (B) is the P_{CO}/P_{CO_2} (T') value at which Me and MeO_2 coexist. At P_{CO}/P_{CO_2} values of lower than P_{CO}/P_{CO_2} (T'), $RT \ln P_{O_{2,e}}$ of the $CO-CO_2$ gas mixture at temperature T' is higher than Point A. Thus, the partial oxygen pressure, $P_{O_{2,e}}$, of the $CO-CO_2$ gas mixture in an equilibrium state (i.e. *equilibrium oxygen partial pressure*) is higher than the $P_{O_{2,e}}$ of MeO_2. Hence, MeO_2 is stable. Similarly, at P_{CO}/P_{CO_2} values exceeding P_{CO}/P_{CO_2} (T'), Me is stable.

(2) Temperature, T', at which Me and MeO_2 coexist for a given P_{CO}/P_{CO_2} value (Point B)

Draw the B–C line and find the point that intersects with the $\Delta G^\circ - T$ line (Point A). Then, draw a vertical line in the direction of the temperature scale. The intersecting point (T') is the temperature value we are looking for.

10.5.3 Equilibria Between H_2–H_2O and MeO_2, S_2 and MeS_2, and H_2–H_2S and MeS_2

As in the CO–CO_2 and MeO_2 scenario, the equilibrium between H_2–H_2O and MeO_2 can be found using an Ellingham diagram.

The standard Gibbs energy change, ΔG°, of the reaction

$$2H_2(g) + O_2(g) = 2H_2O(g) \tag{10.143}$$

is

$$\Delta G^\circ = -RT \ln \frac{P_{H_2O}^2}{P_{H_2}^2 P_{O_2}} \tag{10.144}$$

where P_i is the partial pressure of component i.

Therefore,

$$RT \ln P_{O_2} = \Delta G^\circ - 2RT \ln \frac{P_{H_2}}{P_{H_2O}} \tag{10.145}$$

Hence, the partial oxygen pressure of a H_2–H_2O gas mixture in an equilibrium state is decided by the values of ΔG° and P_{H_2}/P_{H_2O}.

By extrapolating the relationship between $RT \ln P_{O_2}$ and T for $P_{H_2}/P_{H_2O} = 1$ to 0 K, we obtain Point H in Figs. 10.9 and 10.10. The P_{O_2} value on the line between Point H and the point on the P_{H_2}/P_{H_2O} scale is equal to the P_{O_2} value calculated using Eq. (10.145). Therefore, by using Point H and the scale for P_{H_2}/P_{H_2O} instead of Point C and P_{CO}/P_{CO_2} (as in the CO–CO_2 case), the P_{H_2}/P_{H_2O} ratio, i.e. the state of equilibrium between H_2–H_2O and MeO_2, can be obtained.

The dissociation pressure of sulphur can be obtained from an Ellingham diagram in a similar manner to the dissociation pressure of oxygen by considering the reaction

$$Me(s, \ell) + S_2(g) = MeS_2(s, \ell) \tag{10.146}$$

An Ellingham diagram for sulphides is shown in Fig. 10.12.

Points S and H in Fig. 10.12 correspond to Points O and H, respectively, for P_{H_2}/P_{H_2O} in the Ellingham diagram for oxides (Fig. 10.10). In the Ellingham diagram for sulphides (Fig. 10.12), the scales for P_{S_2} and P_{H_2}/P_{H_2S} are shown.

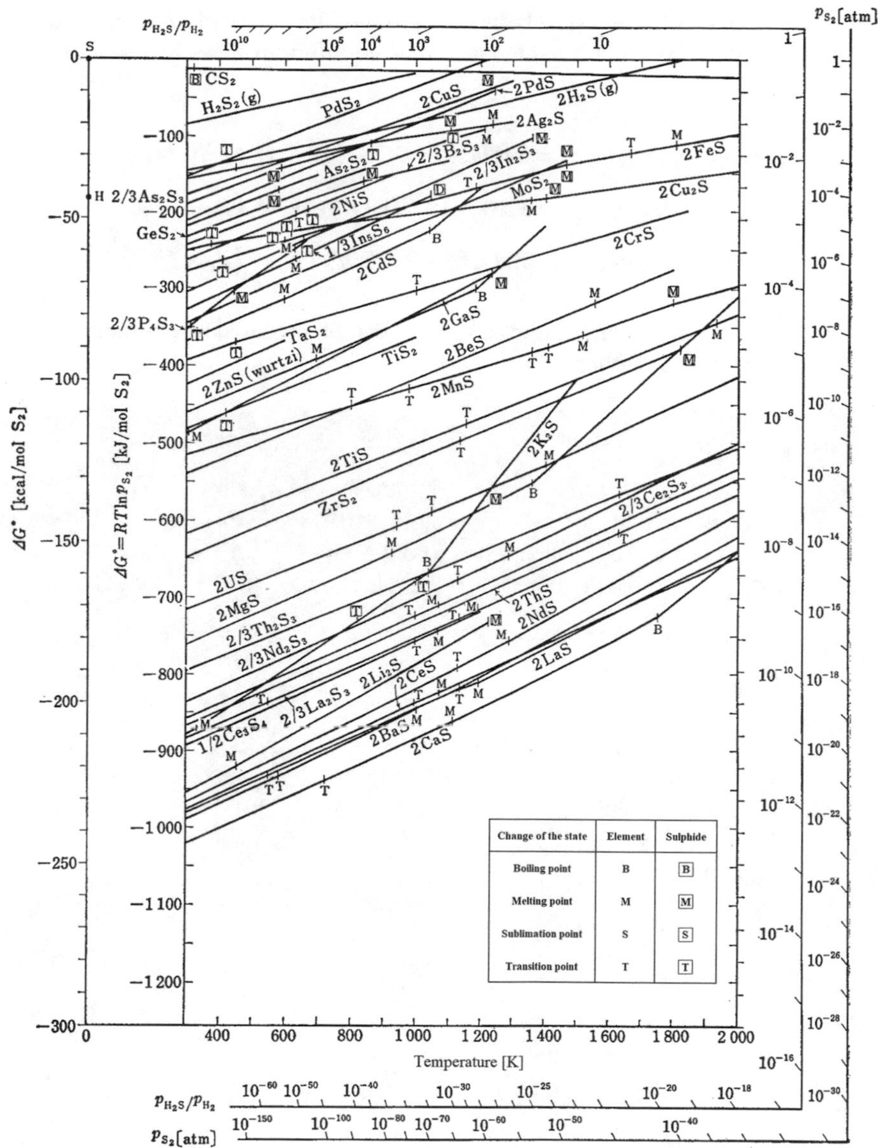

Fig. 10.12 The relationship between the standard Gibbs energy of formation of sulphides and temperature (*Handbook of Iron and Steel, I (Fundamentals)*, 3rd Ed., ed. by ISIJ,. Maruzen, Tokyo, 1981)

Obtaining an equilibrium state between a gas mixture and oxides/sulphides using an Ellingham diagram is equivalent to obtaining equilibrium states for the following reactions:

$$2CO(g) + MeO_2(s, \ell) = Me(s, \ell) + 2CO_2(g) \quad K = \left(P_{CO_2}/P_{CO}\right)^2$$

$$2H_2(g) + MeO_2(s, \ell) = Me(s, \ell) + 2H_2O(g) \quad K = \left(P_{H_2O}/P_{H_2}\right)^2$$

$$2H_2(g) + MeS_2(s, \ell) = Me(s, \ell) + 2H_2S(g) \quad K = \left(P_{H_2S}/P_{H_2}\right)^2$$

The dissociation pressure of nitrogen, $P_{N_{2,e}}$, can be obtained in a similar manner to that of oxygen and sulphur.

Example 10.19

(1) Determine the H_2O concentration (mol%) at which hydrogen gas must be maintained to obtain iron from iron ore at 1073 K (800 °C).

(2) Determine the CO_2 concentration (mol%) at which carbon monoxide gas must be maintained to obtain iron from iron ore at 1073 K (800 °C).

Solutions:

(1)

The reduction of iron ore in the final stage is expressed by

$$FeO(s) + H_2(g) = Fe(s) + H_2O(g)$$

Using Fig. 10.9, when the above reaction is in an equilibrium state at 1073 K (800 °C), $P_{H_2}/P_{H_2O} = 5$.

On the other hand,

$$\frac{x_{H_2}}{x_{H_2O}} = \frac{P_{H_2}}{P_{H_2O}} = 5$$

and

$$x_{H_2} + x_{H_2O} = 1$$

Therefore,

$$x_{H_2O} = 0.17(17\,\text{mol\%})$$

Hence, the H_2O concentration in H_2 must be less than 17 mol% in order for iron ore to be reduced to iron.

(2)

This can be calculated in a similar manner to (1).

To obtain Fe from FeO at 1073 K (800 °C), the CO_2 concentration in CO must be less than 13 mol%.

Example 10.20

(1) Calculate $P_{S_{2,e}}$ at 873 K (600 °C) for Ag_2S, MnS, and CaS.

(2) Calculate P_{H_2S}/P_{H_2} for a H_2–H_2S gas mixture that is in equilibrium with Ag_2S, MnS, and CaS at 873 K (600 °C).

Solutions:

From Fig. 10.12, we can see that at 873 K (600 °C)

$$\text{Ag}_2\text{S} : P_{S_{2,e}} = 4 \times 10^{-7}\text{bar},\ P_{H_2S}/P_{H_2} = 0.7$$
$$\text{MnS} : P_{S_{2,e}} = 3 \times 10^{-27}\text{bar},\ P_{H_2S}/P_{H_2} = 6 \times 10^{-11}$$
$$\text{CaS} : P_{S_{2,e}} = 1 \times 10^{-55}\text{bar},\ P_{H_2S}/P_{H_2} = 4 \times 10^{-25}$$

10.6 Summary

The van't Hoff isotherm:

The equation for calculating the equilibrium position, the *van't Hoff isotherm,* is expressed as follows:

$$\Delta G^\circ = -RT \ln K \tag{10.6}$$

where ΔG° is the standard Gibbs energy change, R is the gas constant, T is the temperature, and K is the equilibrium constant.

Obtaining ΔG°:

ΔG° can be obtained using

(1) The free-energy function

$$\frac{G_T^\circ - H_0^\circ}{T} = \int_0^T (H_T^\circ - H_0^\circ) \mathrm{d}\left(\frac{1}{T}\right) \tag{10.66}$$

(2) The standard Gibbs energy of formation, $\Delta_f G_i^\circ$

$$\Delta G^\circ = \sum_{i=1}^{r} \nu_i \Delta_f G_i^\circ \tag{10.88}$$

(3) Heat capacity

$$\Delta G^\circ = \Delta H_0^{\circ\prime} - \Delta aT \ln T - \frac{1}{2}\Delta b T^2 - \frac{1}{2}\Delta c T^{-1} + \left(\Delta a - \Delta S_0^{\circ\prime}\right)T \tag{10.95}$$

where $\Delta H_0^{\circ\prime}$ and $\Delta S_0^{\circ\prime}$ are integral constants and $\Delta C_P^\circ = \Delta a + \Delta b T + \Delta c T^{-2}$.

The van't Hoff equation:
The equation

$$\Delta H^\circ = \sum_{i=1}^{r} \nu_i \overline{H}_i^\circ \tag{10.127}$$

is the *van't Hoff equation.* ΔH° is the standard enthalpy change, ν_i is the stoichiometric number of the reaction (which is negative for reactants and positive for products), and $\overline{H}_i^\circ$ is the standard partial molar enthalpy of component i. The temperature dependence of the equilibrium constant, K, can be described as follows in the temperature range at which ΔH° is considered to be virtually constant:

$$\ln K = \frac{-\Delta H^\circ}{RT} + \mathrm{I} \tag{10.128}$$

Ellingham diagram:
An Ellingham diagram plots standard Gibbs energy change (Gibbs energy of formation) against temperature in order to ascertain equilibrium temperature, equilibrium pressure, stability of compounds, etc.

Chapter 11
Introduction to Computational Thermodynamics

In this chapter, the correlation between chemical thermodynamics, phase diagrams, and the principles of calculation methods are discussed. The calculations are performed using Gibbs energy, and so accurately assessing its value is very important. Hence, several models for assessing Gibbs energy are discussed in the latter section of this chapter.

11.1 Introduction

Phase diagrams are crucial to understanding the correlation between composition, microstructure, and material properties. Therefore, several phase diagram collections have been published, including:

(1) *Binary Alloy Phase Diagrams,* 2nd Edition, three-volume set. Edited by T. B. Massalski, H. Okamoto, P.R. Subramanian, L. Kacprazak. ASM International (1990).
(2) *Handbook of Ternary Alloy Phase Diagrams*, ten-volume set. Edited by P. Villars, A. Prince, H. Okamoto. ASM International (1995).

However, the number of a diagram, i.e. the number of combinations of elements, is large, even for binary or ternary systems; for multicomponent systems, it is almost impossible to create diagrams by experiments. As a result, the CALculation of PHAse Diagram (CALPHAD) method was developed based on knowledge of thermodynamics.

Several calculation software packages are commercially available, with Thermo-Calc and FactSage being among the better-known. The CALPHAD method is useful when data for phase diagrams, particularly those relating to multicomponent systems, is not available, as it allows phase diagrams to be created without tremendous amounts of time and labour being expended on experiments.

T. Matsushita and K. Mukai, *Chemical Thermodynamics in Materials Science*,
https://doi.org/10.1007/978-981-13-0405-7_11

11.2 Binary Isomorphous Phase Diagrams

For the following equilibrium calculation, Gibbs energy, G, will be used. Under given conditions (e.g. temperature, composition, pressure), the Gibbs energy change, ΔG, is zero in an equilibrium state. This can be understood in relation to the motion of the ball on the slope (equivalent to the stability of a system) in Fig. 11.1: The y axis is the Gibbs energy of the system, and the x axis is the direction of the reaction or path of the state change (corresponding to e.g. the concentration of Component B of an A–B binary system).

As is discussed in Chap. 10, a reaction continues (Fig. 11.1a) until $\Delta G = 0$ is achieved, at which point the system is stable (i.e. in an equilibrium state). In addition to the equilibrium state at the minimum G point shown in Fig. 11.1b, the system may become stable at the local minimum G point shown in Fig. 11.1c. In such a case, the equilibrium is termed *metastable equilibrium*. Furthermore, the system may reach equilibrium at the maximum G point shown in Fig. 11.1d, which is very unstable.

Based on these concepts, we will first consider the calculation principle of one of the simplest forms of phase diagram, an A–B binary isomorphous phase diagram. The Gibbs energy, G, of the liquid and solid phases of the A–B binary system at different temperatures are shown in Fig. 11.2.

At temperature T_1, the Gibbs energy of the liquid phase, G_L, is lower than that of the solid phase, G_S, across the composition range. Hence, as is shown in Fig. 11.3, at temperature T_1, the liquid phase is more stable at equilibrium than the solid phase across the composition range. For temperature T_2, the Gibbs energy curves of the solid and liquid phases intersect (Fig. 11.2), and so it is possible to draw a common tangent line; in the composition range between the tangent points (composition between a and b), the solid and liquid phases can co-exist. Hence, and as shown in Fig. 11.3, the solid and liquid phases co-exist at temperature T_2 between composition a and b. In the same way, the liquid and solid phases co-exist between composition c and d at temperature T_3. At temperature T_4, the Gibbs energy of the liquid phase, G_L, is higher than that of the solid phase, G_S, across the composition range. Hence, and as shown in Fig. 11.3, at temperature T_4 the solid phase is more stable at equilibrium than the liquid phase across the composition range.

Phase diagrams can be calculated using Gibbs energy data at each temperature. The above-mentioned software products utilise this principle, and each company produces and maintains Gibbs energy databases for this purpose.

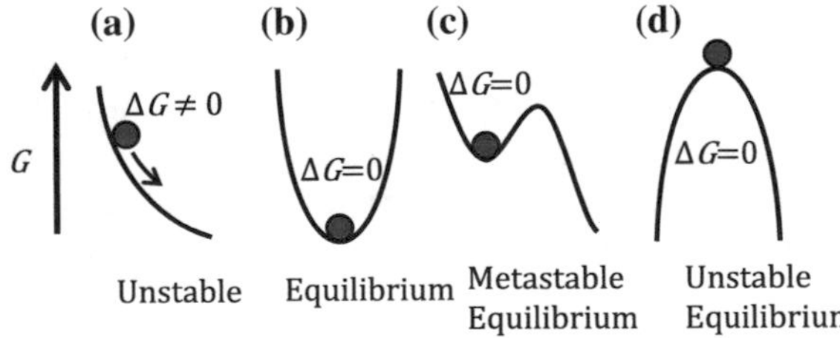

Fig. 11.1 Conceptual illustration of an equilibrium state

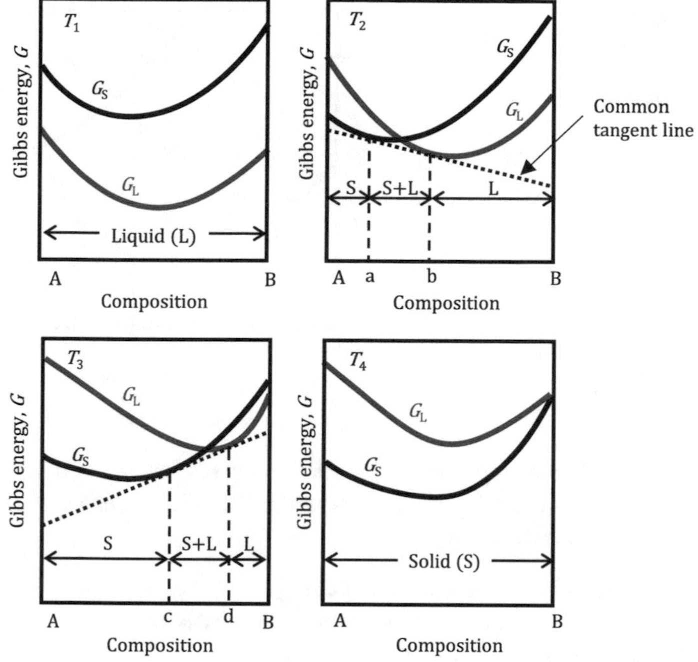

Fig. 11.2 Schematic illustration of the Gibbs energy of an A–B binary system (solid and liquid phases)

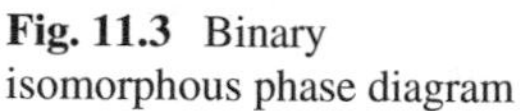

Fig. 11.3 Binary isomorphous phase diagram

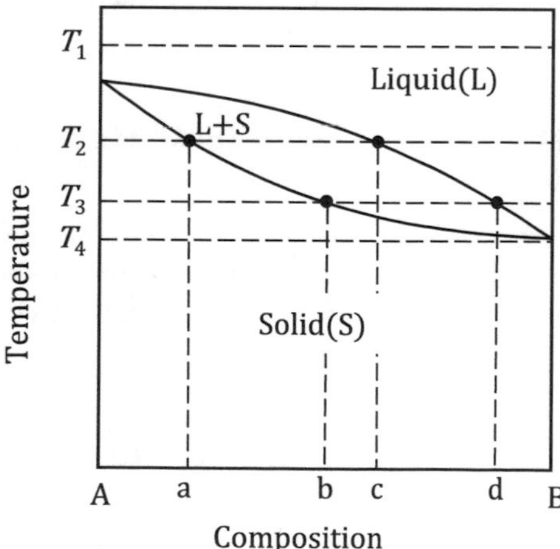

11.3 Binary Systems—Other Types of Phase Diagram

In the previous section, a binary isomorphous phase diagram was taken as an example for the sake of simplicity. The principle can, however, be applied to systems consisting

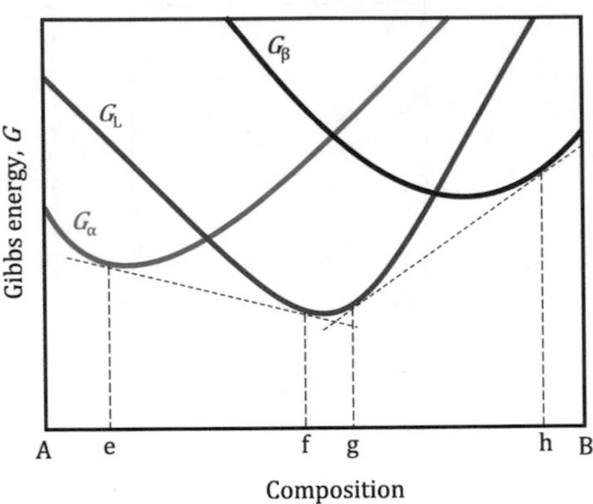

Fig. 11.4 Schematic illustration of the Gibbs energy of an A–B binary system (α, β, and liquid phases)

of different phases. For example, the Gibbs energy curves for α, β, and liquid phases at certain temperatures can be drawn, as shown in Fig. 11.4. In this case, the stable phases at each composition range are:

- Between pure A and composition e: α phase
- Between composition e and f: α phase and liquid phase
- Between composition f and g: liquid phase
- Between composition g and h: β phase and liquid phase
- Between composition h and Pure B: β phase.

In the same way, e.g. eutectic and peritectic phase diagrams can also be calculated.

11.4 Phase Diagrams of Multicomponent Systems

The above-mentioned principles can also be applied to multicomponent systems, although the calculations become much more complex. Thus, thermodynamic software packages such as Thermo-Calc are commonly utilised for this purpose.

11.5 Gibbs Energy of a Regular Solution

As was discussed in previous chapters, Gibbs energy is an important factor in phase diagram calculation. In the following sections we will explore the ways in which Gibbs energy can be described.

In the case of a regular solution (Sect. 8.5.4), the entropy of mixing, $\Delta_{\text{mix}} S$, is the same as that of an ideal solution, but the enthalpy of mixing, $\Delta_{\text{mix}} H$, is not zero (wheras that of an ideal solution is). The Gibbs energy of 1 mol of a regular solution (integral molar Gibbs energy, $\overline{G}$) can be described as follows:

$$\overline{G} = \overline{G}^* + \Delta_{\text{mix}}\overline{H}^{\text{reg}} - T\Delta_{\text{mix}}\overline{S}^{\text{reg}} \tag{11.1}$$

As is stated in Chap. 9, $\overline{G}^*$ is the Gibbs energy of an unreacted A–B mixture, and can be described as $\overline{G}^* = x_A\overline{G}_A^* + x_B\overline{G}_B^*$.

The enthalpy of mixing of 1 mol of a regular solution, $\Delta_{\text{mix}}\overline{H}^{\text{reg}}$, can be described as

$$\Delta_{\text{mix}}\overline{H}^{\text{reg}} = \Omega x_A x_B \tag{11.2}$$

where Ω is the interaction parameter and x_i is the mole fraction of component i.

The interaction parameter Ω can be either positive or negative. Therefore, the term $\Omega x_A x_B$ can also be either positive or negative. The shape of a Gibbs energy curve, and thus the phase diagram itself, is affected by the sign of Ω.

When Ω is negative, for example—i.e. the enthalpy of mixing, $\Delta_{\text{mix}}\overline{H}^{\text{reg}}$, is negative (as shown in Fig. 11.5a)—the Gibbs energy curve has a U shape. When Ω is positive, the Gibbs energy curve appears as in Fig. 11.5b (i.e. it has two inflection points when the temperature, T, is lower than $\Omega/2R$, as is discussed below).

The entropy term of a regular solution ($T\Delta_{\text{mix}}\overline{S}^{\text{reg}}$) can be described as follows (see Eq. (6.31)):

$$T\Delta_{\text{mix}}\overline{S}^{\text{reg}} = -RT(x_A \ln x_A + x_B \ln x_B) \tag{11.3}$$

where $\Delta_{\text{mix}}\overline{S}^{\text{reg}}$ is the entropy of mixing of 1 mol of a regular solution.

Hence, the Gibbs energy of 1 mol of a regular solution can be described as follows:

$$\overline{G} = x_A\overline{G}_A^* + x_B\overline{G}_B^* + \Omega x_A x_B + RT(x_A \ln x_A + x_B \ln x_B) \tag{11.4}$$

It should be noted that an A–B system tends to separate into two phases when the interaction parameter, Ω, is positive due to the fact that the A atoms tend to form pairs with other A atoms and repel B atoms, and vice versa. Hence, the Gibbs energy curve ($\overline{G}$) has an undulating shape, as shown in Fig. 11.5b. As a result, Eq. (11.4) may be rewritten as follows:

$$\begin{aligned}\overline{G} =& x_B(\overline{G}_B^* - \overline{G}_A^*) + \overline{G}_A^* + \Omega x_B(1 - x_B)\\ &+ RT[(1 - x_B)\ln(1 - x_B) + x_B \ln x_B]\end{aligned} \tag{11.5}$$

as

$$x_A\overline{G}_A^* + x_B\overline{G}_B^* = x_B(\overline{G}_B^* - \overline{G}_A^*) + \overline{G}_A^*$$

$$\Omega x_A x_B = \Omega x_B(1 - x_B)$$

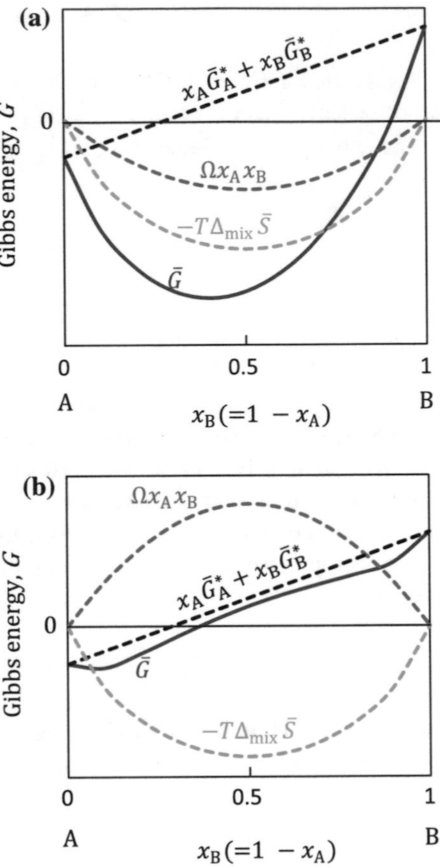

Fig. 11.5 The influence of the interaction parameter Ω on the Gibbs energy curve

$$RT(x_A \ln x_A + x_B \ln x_B) = RT[(1 - x_B)\ln(1 - x_B) + x_B \ln x_B]$$

The second-order partial derivatives of Eq. (11.5) are

$$\frac{\partial^2 \overline{G}}{\partial x_B^2} = \frac{RT}{x_B} + \frac{RT}{1 - x_B} - 2\Omega \tag{11.6}$$

The Gibbs energy curve in Fig. 11.5b has inflection points at the points at which $\frac{\partial^2 \overline{G}}{\partial x_B^2} = 0$. The solution of

$$\frac{RT}{x_B} + \frac{RT}{1 - x_B} - 2\Omega = 0 \tag{11.7}$$

is

$$x_B = -\frac{\sqrt{\Omega^2 - 2\Omega RT} - \Omega}{2\Omega},\ x_B = \frac{\sqrt{\Omega^2 - 2\Omega RT} + \Omega}{2\Omega}$$

Hence, when the temperature, T, is lower than $\Omega/2R$, the Gibbs energy curve in Fig. 11.5b has two inflection points and an undulating shape. When T is higher than $\Omega/2R$, the Gibbs energy curve is at its minimum at the point at which the first-order partial derivative, $\frac{\partial \overline{G}}{\partial x_B}$, is zero.

11.6 General Form of Gibbs Energy

In a general form, integral molar Gibbs energy can be described as follows:

$$\overline{G} = \overline{G}^* - T\Delta_{\text{mix}}\overline{S} + \overline{G}^{\text{Phys}} + \overline{G}^{\text{E}} \tag{11.8}$$

$\overline{G}^*$ is the Gibbs energy of an unreacted mixture (of 1 mol in total) and, in the case of a binary system, can be described as $\overline{G}^* = x_A G_A^* + x_B G_B^*$.

$-T\Delta_{\text{mix}}\overline{S}$ is the entropy term, and is a function of the number of possible arrangements (microscopic states) of the system (constituents in the phase), $\mathcal{W}$ (see Sect. 6.1). In the case of ideal or regular binary solutions, $-T\Delta_{\text{mix}}\overline{S}$ is described by $RT(x_A \ln x_A + x_B \ln x_B)$ (see Eq. (6.31)).

$\overline{G}^{\text{Phys}}$ is the contribution of physical phenomena such as magnetic transitions to Gibbs energy.

$\overline{G}^{\text{E}}$ is the excess molar Gibbs energy (see Sect. 9.2), i.e. the total Gibbs energy that is not included in the term $\overline{G}^* - T\Delta_{\text{mix}}\overline{S} + \overline{G}^{\text{Phys}}$. In the case of a regular binary solution, $\overline{G}^{\text{E}} = \Delta_{\text{mix}}\overline{H} = \Omega x_A x_B$, as $\Delta_{\text{mix}}\overline{H}_i = RT\ \ln f_i = \overline{G}_i^{\text{E}}$.

The accuracy of phase diagram calculation results depends on the accuracy of the Gibbs energy values used for the calculation. It is therefore important to assess Gibbs energies as accurately as possible, and so several models for describing Gibbs energy have been suggested, including the *sublattice model*.

11.7 The Sublattice Model

In Thermo-Calc, the sublattice model is used to describe the Gibbs energy of an ordered phase or disordered system that includes interstitial solute elements. Figure 11.6 shows an example of a sublattice, which has an ordered bcc structure and is classified as B2 in *Strukturbericht* and cP2 in *Pearson symbol* (*Pearson notation*).

The lattice consists of Sublattice 1 (in which A atoms are dominant) and Sublattice 2 (in which B atoms are dominant). In the case of a Compound AB that has a B2

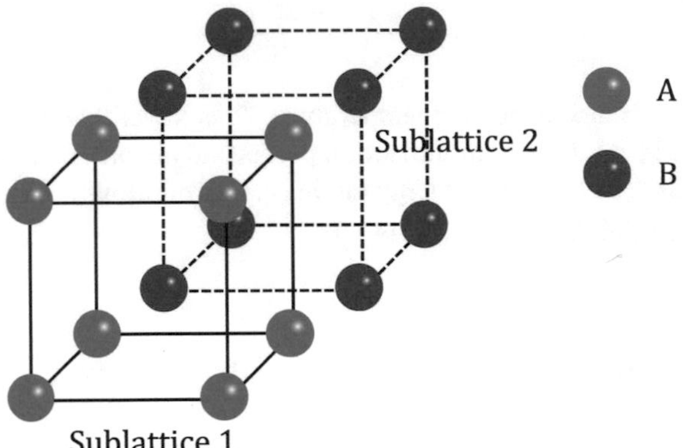

Fig. 11.6 Sublattice

structure, the A atoms can also occupy Sublattice 2 and vice versa, and it is denoted by (A,B)(B,A).

The description of Gibbs energy depends on the number of sublattices but, in the case of Compound AB, is as follows:

$$\begin{aligned} G = & x_A^{(1)} x_A^{(2)} G_{A:A}^* + x_B^{(1)} x_B^{(2)} G_{B:B}^* + x_A^{(1)} x_B^{(2)} G_{A:B}^* + x_B^{(1)} x_A^{(2)} G_{B:A}^* \\ & + RT\left(x_A^{(1)} \ln x_A^{(1)} + x_B^{(1)} \ln x_B^{(1)} + x_A^{(2)} \ln x_A^{(2)} + x_B^{(2)} \ln x_B^{(2)}\right) \\ & + x_A^{(1)} x_B^{(1)} x_A^{(2)} L_{A,B:A} + x_A^{(1)} x_B^{(1)} x_B^{(2)} L_{A,B:B} + x_A^{(1)} x_A^{(2)} x_B^{(2)} L_{A:A,B} \\ & + x_B^{(1)} x_A^{(2)} x_B^{(2)} L_{B:A,B} \end{aligned} \tag{11.9}$$

where $x_A^{(1)}$ is the mole fraction of Atom A in Sublattice 1, and $G_{A:B}^*$ is the Gibbs energy of AB when Sublattice 1 is occupied only by A atoms and Sublattice 2 is occupied only by B atoms. The colon between X and Y ('X:Y') indicates that X and Y exist on different sublattices. $G_{A:A}^*$ is the Gibbs energy when both Sublattice 1 and Sublattice 2 are occupied by A atoms. $L_{i,j:k}$ is an interaction parameter, and relates to the free energy of mixing of Atoms i and j when Sublattice 1 is occupied by Atoms i and j and Sublattice 2 is occupied by Atom k. $L_{k:i,j}$ is an interaction parameter, and relates to the free energy of mixing of Atoms i and j when Sublattice 1 is occupied by Atom k and Sublattice 2 is occupied by Atoms i and j.

For a regular solution model, there are no temperature and composition dependences on the interaction parameter Ω. However, for a real system (e.g. an alloy system), temperature and composition dependences exist for the interaction parameter Ω, whereas these dependences are taken into account in order to calculate phase diagrams more precisely by introducing the interaction parameter L.

The interaction parameter L can be described using the *Redlich-Kister power series*:

$$L_{ij} = \sum_{\nu=0}^{k} (x_i - x_j)^{\nu} \cdot {}^{\nu}L_{ij} \tag{11.10}$$

The value of Ω for the regular soution model corresponds to L_{ij} in the above equation when $\nu = 0$, i.e.${}^{0}L_{ij}$

If necessary, a higher order term can be introduced. However, in most cases $\nu = 2$ is sufficient to describe the interactions between atoms.

11.8 Multicomponent Systems

In the previous section, the explanation of Gibbs energy was limited to binary systems in the interest of simplicity; here, it will be expanded to include multicomponent systems. For this purpose, the excess term of the ternary interaction contribution,$G^{\mathrm{E,Ter}}$, and the excess term of the higher-order interaction contribution, $G^{\mathrm{E,Higher}}$, should be added to the excess term G^{E} in Eq. (11.8). Hence, G^{E} in Eq. (11.8) may be expressed as follows for a multicomponent system:

$$G^{\mathrm{E}} = G^{\mathrm{E,Bin}} + G^{\mathrm{E,Ter}} + \cdots + G^{\mathrm{E,Higher}} \tag{11.11}$$

$G^{\mathrm{E,Ter}}$ can be described as follows:

$$G^{\mathrm{E,Ter}} = \sum_{i=1}^{n-2} \sum_{j=i+1}^{n-1} \sum_{k=j+1}^{n} x_i x_j x_k L_{ijk} \tag{11.12}$$

For a higher-order interaction contribution, $G^{\mathrm{E,Higher}}$ can be described as follows:

$$G^{\mathrm{E,Higher}} = \sum_{i=1}^{n-3} \sum_{j=i+1}^{n-2} \sum_{k=j+1}^{n-1} \sum_{l=k+1}^{n} x_i x_j x_k x_l L_{ijkl} + \cdots \tag{11.13}$$

11.9 Other Models

Several models to describe the influence of e.g. pressure and magnetism on Gibbs energy have been suggested. For example, according to the Murnaghan model, the pressure dependence of integral molar Gibbs energy can be described as follows:

$$\begin{aligned} \overline{G}(T, P) = {} & \overline{G}(T, P = 0) \\ & + \frac{V_0 \exp\left(\int_{298}^{T} \alpha(T)\mathrm{d}T\right)}{(\mathrm{n} - 1)K_0(T)} \left[(1 + \mathrm{n}K_0(T)P)^{1-1/\mathrm{n}} - 1\right] \end{aligned} \tag{11.14}$$

where T is the temperature, P is the pressure, V_0 is the volume at zero pressure, $\alpha(T)$ is the thermal expansivity, $K_0(T)$ is the compressibility at zero pressure, and n is a constant that is independent of temperature and pressure.

In addition, according to a model (Lukas et al. 2007), the pressure dependence of integral molar Gibbs energy can be described as follows:

$$\overline{G}(T,P) = \overline{G}(T,P=0) + \frac{c(T)}{\kappa_0(T)}\left[\exp\left(\frac{V(T,P)-V_0(T)}{c(T)}\right) - 1\right] \quad (11.15)$$

where $c(T)$ is termed the *adjustable function*,$\kappa_0(T)$ is the compressibility at zero pressure, $V(T,P)$ is the volume at temperature T and pressure P, and $V_0(T)$ is the volume at zero pressure.

Regarding the influence of magnetism on Gibbs energy, the*Hillert-Jarl model* is widely accepted; here, the contribution of magnetic phenomena to Gibbs energy, G^{Mag}, can be described as follows:

$$G^{\mathrm{Mag}} = nRTf(\tau)\ln(\beta+1) \quad (11.16)$$

where n is the number of atoms per formula unit that have the average magnetic moment β, R is the gas constant, T is the temperature, and

when $\tau < 1$,

$$f(\tau) = 1 - \frac{1}{A}\frac{79\tau^{-1}}{140p} + \frac{474}{497}\left(\frac{1}{p}-1\right)\frac{\tau^3}{6} + \frac{\tau^9}{135} + \frac{\tau^{15}}{600} \quad (11.17)$$

When $\tau \geq 1$

$$f(\tau) = -\frac{1}{A}\left(\frac{\tau^{-5}}{10} + \frac{\tau^{-15}}{315} + \frac{\tau^{-25}}{1500}\right) \quad (11.18)$$

where

$$A = \frac{518}{1125} + \frac{11692}{15975}\left(\frac{1}{p}-1\right)$$

p depends on the structure and is termed the *structure factor*, $\tau = T/T_{\mathrm{C}}$, and T_{C} is the Curie temperature.

11.10 Summary

In this chapter, the principle of calculating phase diagrams using the Gibbs energy of different phases has been explained in brief, and several models for describing

Gibbs energy have been introduced. A detailed discussion of Gibbs energy models can be found in Lukas et al. (2007).

In this book, phase diagram calculations have been discussed. However, the CALPHAD method facilitates e.g. the calculation of thermodynamic functions (using Gibbs energy), simulations of steel-refining processes, the calculation of the Scheil model, and much more. The details of these calculations can be found in the manuals of thermodynamic calculation software packages (see Sect. 12.3).

Reference

Lukas HL, Fries SG, Sundman B (2007) Computational thermodynamics. Cambridge University Press

Chapter 12
Books, Databases, and Software

12.1 Further Reading (in Roughly Ascending Order of Difficulty)

Chemical thermodynamics in general:

- G. Hargreaves, Elementary chemical thermodynamics (3rd Revised edition), Butterworth & Co Publishers Ltd (1973)
- B. Smith, Basic Chemical Thermodynamics (6th edition), Imperial College Press (2013)
- G. K. Vemulapalli, Physical Chemistry, Prentice-Hall (1993)
- D. H. Everett, An Introduction to the Study of Chemical Thermodynamics (Second edition), Longman (1971)
- J. G. Kirkwood and I. Oppenheim, Chemical thermodynamics, McGraw-Hill (1961)
- I. Prigogine and R. Defay, Chemical thermodynamics (Translated by D. H. Everett), Longmans Green and Co. (1954)

Chemical thermodynamics in relation to materials science and metallurgy:

- O. Kubaschewski and C. B. Alcock, Metallurgical Thermochemistry (Fifth edition), Pergamon press (1979)
- Hae-Geon Lee, Materials Thermodynamics: With Emphasis on Chemical Approach, World Scientific Publishing Co (2012)
- D. R. Gaskell, Introduction to the Thermodynamics of Materials (Fifth edition), CRC press (2008)
- C. H. P. Lupis, Chemical Thermodynamics of Materials, North-Holland (1983)
- L. S. Darken and R. W. Gurry, Physical Chemistry of Metals, McGraw-Hill book Co. (1953)

T. Matsushita and K. Mukai, *Chemical Thermodynamics in Materials Science*,
https://doi.org/10.1007/978-981-13-0405-7_12

Computational chemical thermodynamics:

- M. Hillert, Phase Equilibria, Phase Diagrams and Phase Transformations: Their Thermodynamic Basis (Second Edition), Cambridge University Press (2008)
- H. L. Lukas, S. G. Fries and B. Sundman, Computational thermodynamics: The Calphad Method, Cambridge University Press (2007)

12.2 Thermodynamic Data Books

12.2.1 Data Books

If the standard Gibbs energy change, ΔG°, of a reaction is known, the equilibrium constant, K, or direction of the reaction can be easily obtained using the van't Hoff isotherm. Hence, it is important that ΔG° values are reliable. In this section, several thermodynamic data books that give values for ΔG°, ΔH°, C_P°,S°, $H^\circ - H_{298}^\circ$, $\Delta_f H^\circ$, $\Delta_f G^\circ$, free-energy function, etc. are listed.

It is important to check the meaning of symbols, the units (atm or bar, J or cal, etc.), the legend, and so on as symbols differ from book to book. The symbol F is particularly confusing as it is used for Helmholtz energy in some books and Gibbs energy in others. Furthermore, one must distinguish between ΔG° and G_i°; for G_i°, we must check the standard state due to the fact that the values vary with standard state. Sometimes the same symbol is used to mean something different, or the same quantity is denoted by different symbols.

Once again; it is important to check the definition of each symbol and its units (Kelvin or Celsius; cal, kcal, J, or kJ; per mole, per gram, or per kg, etc.).

12.2.2 General Materials/Substances

Below is a list of thermodynamic data books for general materials:

(1) JANAF Thermochemical Tables, 2nd ed., D. R. Stull and H. Praphet, 1971, NSRDS. The latest version of is published online. http://kinetics.nist.gov/janaf/
(2) Thermochemical Properties of Inorganic Substances, I. Barin and O. Knacke, 1973, Springer-Verlag and I. Barin, O. Knacke and O. Kubaschewski, 1977, Supplements
(3) Selected Values of the Thermodynamic Properties of the Elements, R. Hultgren et al., 1973, ASM
(4) Thermodynamic Data for Inorganic Sulphides, Selenides and Tellurides, K. C. Mills, 1974, Butterworths
(5) Thermochemical Data of Pure Substances, Part I, II, I. Barin, 1989, VCH Publishers

12.2.3 Metals and Oxides

The thermodynamic data for metals and oxides can be found in the following data books:

(1) Thermochemistry for Steelmaking, Vol. I, II., J. F. Elliott et al., 1963, Addison-Wesley
(2) Selected values of the Thermodynamic Properties of Binary Alloys, R. Hultgren et al., 1973, ASM
(3) Metallurgical Thermochemistry, 5th ed., O. Kubaschewski and C. B. Alcock, 1979, Pergamon Press
(4) Steelmaking Data Sourcebook: The Japan Society for Promotion of Science: The 19th Committee on Steelmaking, New York, Gordon and Breach Science Publishers, 1988

12.3 Thermodynamic Databases and Software

In addition to the above-mentioned classical data books, today thermodynamic data is available as part of thermodynamic calculation software packages. The Gibbs energies of elements, components, and phases are provided as TDB (Thermodynamic DataBase) format files, which were originally designed for use with Thermo-Calc but have since become the de facto standard database format used by thermodynamic calculation software packages.

Several major thermodynamic calculation software packages and their websites are listed below:

(1) Thermo-Calc: http://www.thermocalc.com/
(2) FactSage: http://www.factsage.com/
(3) OpenCalphad: http://www.opencalphad.com/
(4) MatCalc: http://matcalc.tuwien.ac.at/
(5) PANDAT: http://www.computherm.com/

Most companies provide databases for each material (e.g. iron alloys, oxides, slag) in TDB format. Although most database files restrict users' access to parameter values, several free database files exist that allow users full access:

(1) SGTE UNARY database (database for unary systems): http://www.crct.polymtl.ca/sgte/index.php?free=1
(2) Computational Phase Diagram Database (a database of approximately 250 binary and higher-order systems): http://cpddb.nims.go.jp/index_en.html
(3) Steel, Ni base alloy, Al base alloy database: http://matcalc.tuwien.ac.at/index.php/databases/open-databases

Chapter 13
Thermodynamic Data

13.1 Standard Gibbs Energy of Formation (Data as a Function of Temperature)

The standard Gibbs energies of formation are summarised in Table 13.1 as a function of temperature. The data is taken from *Handbook of Iron and Steel, I (Fundamentals)*, 3rd Ed., ed. by ISIJ,. Maruzen, Tokyo (1981), and was compiled based on that of Barin and Knacke (1973) and Barin et al. (1977). Note that the temperature ranges in the table are essentially *un*related to phase transformations and any other phenomena, and are intended to be used solely to determine fitting range.

Similar data can be found in Kubaschewski and Alcock (1979) and Turkdogan (1980); the data in the latter was compiled using the following references:

- D. R. Stull and H. Prophet, "JANAF Thermochemical tables," NSRDS-NBS37. U.S. Dept. Commer., Washington, D. C., 1971
- M. W. Chase, J. L. Curnutt, A. T. Hu, H. Prophet, A. N. Syverud, and L. C. Walker, J. Phys. Chem. Ref. Data 3, 311 (1974)
- M. W. Chase, J. L. Curnutt, R. A. McDonald, and A. N. Syverud, J. Phys. Chem. Ref. Data 7, 793 (1978)
- I. Barin and O. Knacke, "Thermochemical Properties of Inorganic Substances." Springer-Verlag, Berlin and New York (1973); Metall. Trans. 5 1769 (1974)
- I. Barin, O. Knacke, and O. Kubaschewski, "Thermochemical Properties of Inorganic Substances, Supplement." Springer-Verlag, Berlin and New York (1977)
- O. Kuwaschewski, E. L. Evans, and C. B. Alcock, "Metallurgical Thermochemistry.", Pergamon, Oxford (1967)
- H. Hultgren, P. D. Desai, D. T. Hawkins, M. Gleiser, K. K. Kelley, and D. D. Wagman, "Selected Values of the Thermodynamic Properties of the Elements." A,. Soc. Met., Metals Park, Ohio (1973)

The equations for the standard Gibbs energy of formation as a function of temperature have been compiled for the reader's convenience. For more accurate values, please refer to the source books or original papers.

T. Matsushita and K. Mukai, *Chemical Thermodynamics in Materials Science*,
https://doi.org/10.1007/978-981-13-0405-7_13

Table 13.1 Standard Gibbs energy of formation, $\Delta_f G_i^\circ = A + BT$ kJ.

Compound	A	B	Temperature range [K]
2 $Al_2O(g)$	−385.76	−0.06658	1500–2000
2 $AlO(g)$	126.58	−0.11478	1500–2000
$Al_2O_2(g)$	−455.83	0.05362	1500–2000
2/3 Al_2O_3	−1121.94	0.21630	1500–2200
AlO_2	−202.67	0.00758	1500–2000
2/3 B_2O_3	−814.08	0.13871	1500–2200
2/3 B_2O_3 (amorphous)	−813.77	0.13811	1500–2200
2 BaO	−1126.36	0.20972	1500–2171
	−1397.90	0.33480	2171–2198
2 BeO	−1191.02	0.18958	1500–1556
	−1210.59	0.20227	1556–2200
2/3 Bi_2O_3	−288.92	0.10143	1500–1800
2 CO(g)	−235.60	−0.16837	1500–2200
$CO_2(g)$	−397.14	−0.00045	1500–2200
2 CaO	−1275.48	0.21131	1500–1757
	−1575.00	0.38194	1757–2200
2 $Cl_2O(g)$	182.63	0.10788	1500–2000
2 ClO(g)	205.61	−0.02863	1500–2000
2 CoO	−468.70	0.14009	1500–1768
	−497.93	0.15661	1768–2078
2/3 Cr_2O_3	−745.48	0.16894	1500–1800
2 Cu_2O	−368.92	0.16594	1500–1509
	−246.70	0.08518	1509–2000
2/3 Dy_2O_3	−1220.76	0.17594	1500–1657
	−1214.51	0.17222	1657–1682
	−1239.31	0.18702	1622–2000
2/3 Er_2O_3	−1245.69	0.17758	1500–1795
	−1267.51	0.18972	1795–2000
2 FeO(g)	−32.09	0.11228	1500–2000
2 FeO	−536.38	0.12344	1500–1650
	−489.38	0.09494	1650–1665
	−486.95	0.09348	1665–1809
	−512.48	0.10762	1809–2200
1/2 Fe_3O_4	−545.47	0.15068	1500–1665
	−545.91	0.15095	1665–1809
	−566.90	0.16254	1809–1870
	−497.72	0.12556	1870–2000

(continued)

Table 13.1 (continued)

Compound	A	B	Temperature range [K]
2/3 Fe_2O_3	−536.63	0.16274	1500–1665
	−536.91	0.16288	1665–1735
2/3 Ga_2O_3	−713.69	0.20815	1500–2068
2/3 Gd_2O_3	−1200.08	0.18245	1500–1533
	−1206.05	0.18638	1533–1585
	−1218.78	0.19449	1585–2200
2 GeO(g)	−157.21	0.09436	1500–2000
2 H_2O(g)	−503.50	0.11611	1500–2200
HfO_2	−1094.28	0.16650	1500–1973
	−1082.85	0.16064	1973–2013
	−1082.87	0.16065	2013–2200
2 HgO(g)	−41.64	0.08139	1500–2000
2/3 In_2O_3	−605.82	0.20243	1500–1523
	−602.73	0.20058	1523–2183
2/3 IrO_3(g)	11.57	0.02983	1500–1800
2 KO(g)	−91.43	0.05440	1500–2000
2 Li_2O	−1190.85	0.26050	1500–1620
	−1767.34	0.61249	1620–1843
	−1633.54	0.53993	1843–2000
2 Li_2O(g)	−384.82	−0.05803	1500–1620
	−974.62	0.30210	1620–2000
2 LiO(g)	140.07	−0.11623	1500–1620
	−156.38	0.06474	1620–2000
Li_2O_2(g)	−261.39	0.01909	1500–1620
	−555.92	0.19892	1620–2000
2 MgO	−1461.10	0.41027	1500–2000
2 MnO	−773.96	0.14979	1500–1517
	−807.89	0.17212	1517–2058
	−701.72	0.12056	2058–2200
1/2 Mn_3O_4	−685.74	0.16783	1500–1517
	−704.79	0.18040	1517–1833
2 MoO(g)	746.51	−0.18261	1500–2000
	734.40	−0.17645	2000–2200
MoO_2	−560.23	0.15642	1500–2000
MoO_2(g)	−24.97	−0.03003	1500–2000
	−30.55	−0.02719	2000–2200
2/3 MoO_3(g)	−242.76	0.04018	1500–2000
	−245.17	0.04139	2000–2200

(continued)

Table 13.1 (continued)

Compound	A	B	Temperature range [K]
2 $N_2O(g)$	176.57	0.13935	1500–2200
2 $NO(g)$	181.92	−0.02624	1500–2200
2/3 $N_2O_3(g)$	62.85	0.12053	1500–2200
$NO_2(g)$	33.73	0.06244	1500–2200
1/2 $N_2O_4(g)$	15.72	0.13919	1500–2200
2/5 $N_2O_5(g)$	15.68	0.13162	1500–2000
2/3 $NO_3(g)$	53.72	0.09601	1500–2000
2 Na_2O	−1044.78	0.47174	1500–2000
2 $NaO(g)$	−48.10	0.05415	1500–2000
2 NbO	−797.11	0.15659	1500–2200
2/5 Nb_2O_5	−744.89	0.16053	1500–1785
	−695.18	0.13274	1785–2200
NbO_2	−766.45	0.15587	1500–2200
2/3 Nd_2O_3	−1209.47	0.19283	1500–2200
2 NiO	−465.74	0.16646	1500–1726
	−499.03	0.18575	1726–2200
2 $PO(g)$	−155.06	−0.02346	1500–2000
1/3 $P_4O_6(g)$	−816.53	0.21706	1500–2000
$PO_2(g)$	−384.54	0.05934	1500–2000
1/5 $P_4O_{10}(g)$ dimeric	−623.82	0.19723	1500–2000
2 PbO (red, ye)	−358.54	0.13589	1500–1700
	−353.34	0.13281	1700–1808
2 $PbO(g)$	62.13	−0.09808	1500–1700
	57.38	−0.09530	1700–2000
2 $PtO_2(g)$	156.77	0.00485	1500–2000
PuO_2	−1040.90	0.17465	1500–2200
2/3 $RuO_3(g)$	−52.80	0.04083	1500–1600
	−53.69	0.04139	1600–1900
1/2 $RuO_4(g)$	−89.12	0.07040	1500–1600
	−89.31	0.07051	1600–2000
2 $S_2O(g)$	−335.94	0.12548	1500–2000
2 $SO(g)$	−115.51	−0.01042	1500–2000
$SO_2(g)$	−360.45	0.07218	1500–1800
2/3 $SO_3(g)$	−303.03	0.10786	1500–2000
2/3 Sb_2O_3	−432.02	0.12644	1500–1729

(continued)

Table 13.1 (continued)

Compound	A	B	Temperature range [K]
2/3 Sc_2O_3	−1265.91	0.19191	1500–1608
	−1271.96	0.19568	1608–1812
	−1291.66	0.20656	1812–2200
2 SeO(g)	−18.95	0.00702	1500–2000
SeO_2	−177.89	0.06604	1500–2000
2 SiO(g)	−220.74	−0.15652	1500–1685
	−325.78	−0.09420	1685–2000
SiO_2 (quart)	−900.06	0.17048	1500–1685
	−923.52	0.18446	1685–1696
SiO_2 (crist.)	−898.45	0.16896	1500–1685
	−946.77	0.19765	1685–1996
	−933.68	0.19108	1996–2200
2/3 Sm_2O_3	−1232.77	0.20091	1500–2000
2 SnO(g)	0.04	−0.09416	1500–2000
SnO_2	−560.60	0.18946	1500–1903
2 SrO	−1188.12	0.20070	1500–1650
	−1471.98	0.37275	1650–1800
2 TaO(g)	414.38	−0.17688	1500–1700
	406.47	−0.17234	1700 2200
TaO_2(g)	−199.48	−0.00898	1500–1700
	−200.77	−0.00824	1700–2200
2/5 Ta_2O_5	−799.71	0.15922	1500–1700
	−793.44	0.15561	1700–2150
	−727.73	0.12498	2150–2200
2 TeO(g)	−14.88	−0.01164	1500–2000
Te_2O_2(g)	−253.32	0.12056	1500–2000
2 ThO(g)	−94.05	−0.12252	1500–1636
	−109.04	−0.11347	1636–2028
	−151.08	−0.09225	2028–2200
ThO_2	−1218.28	0.18086	1500–1636
	−1220.35	0.18213	1636–2028
	−1235.05	0.18938	2028–2200
2 TiO	−1008.72	0.13497	1500–1933
	−1036.74	0.14936	1933–2023
	−925.43	0.09433	2023–2200
2 TiO(g)	−6.39	−0.16195	1500–1933
	−54.29	−0.13711	1933–2200

(continued)

Table 13.1 (continued)

Compound	A	B	Temperature range [K]
2/3 Ti_2O_3	−990.33	0.16372	1500–1933
	−1008.79	0.17322	1933–2112
	−932.52	0.13712	2112–2200
2/5 Ti_3O_5	−966.48	0.16337	1500–1933
	−983.04	0.17191	1933–2047
	−926.45	0.14429	2047–2200
TiO_2 (rutil)	−937.86	0.17692	1500–1933
	−952.76	0.18460	1933–2143
	−883.27	0.15217	2143–2200
TiO_2 (anata)	−927.27	0.17637	1500–1933
	−942.76	0.18440	1933–2000
2/3 Tm_2O_3	−1231.66	0.17610	1500–1800
UO_2	−1086.30	0.17190	1500–2200
2 VO	−829.15	0.14732	1500–1973
2/3 V_2O_3	−792.88	0.14972	1500–2175
	−816.56	0.16052	2175–2200
VO_2	−695.70	0.14780	1500–1633
	−631.13	0.10844	1633–2175
	−646.46	0.11537	2175–2200
2/5 V_2O_5	−578.87	0.12663	1500–2175
	−596.27	0.13460	2175–2200
2 WO(g)	824.30	−0.19077	1500–2000
WO_2	−567.14	0.16287	1500–1997
WO_2(g)	66.15	−0.03522	1500–2000
1/4 W_3O_8(g)	−417.90	0.09342	1500–2000
2/3 WO_3	−544.30	0.15543	1500–1745
	−488.04	0.12325	1745–2110
2/3 WO_3(g)	−196.44	0.03751	1500–2000
1/3 W_2O_6(g)	−378.49	0.07536	1500–2200
2/9 W_3O_9(g)	−439.13	0.10106	1500–2000
1/6 W_4O_{12}(g)	−455.18	0.11012	1500–2000
2/3 Y_2O_3	−1259.70	0.18339	1500–1752
	−1264.07	0.18589	1752–1799
	−1281.71	0.19572	1799–2200
2/3 Yb_2O_3	−1366.69	0.29241	1500–1800
2 ZnO	−915.25	0.39351	1500–2000

(continued)

Table 13.1 (continued)

Compound	A	B	Temperature range [K]
2 ZrO(g)	93.02	−0.13886	1500–2000
	94.59	−0.13967	2000–9125
	52.21	−0.11974	2125–2200
ZrO_2	−1082.53	0.17793	1500–2125
	−1099.22	0.18578	2125–2200
2 AlN	−657.81	0.23447	1500–2000
Be_3N_2	−591.64	0.18741	1500–1556
	−616.68	0.20342	1556–2200
2 CN(g)	920.41	−0.19670	1500–2000
C_2N_2(g)	312.29	−0.04507	1500–2000
2 CeN	−667.14	0.25439	1500–2000
2 GaN	−219.73	0.24167	1500–1773
2/3 NH_3(g)	−113.59	0.23823	1500–1800
2 HfN	−725.59	0.16192	1500–1700
2 LaN	−607.59	0.22160	1500–1800
2 Li_3N	−362.12	0.29109	1500–1620
	−1213.25	0.81119	1620–2000
2 Nb_2N	−486.71	0.15700	1500–2200
2 NbN	−450.86	0.15029	1500–1643
	−433.62	0.13996	1643–2200
2 PN(g)	31.92	−0.01292	1500–2000
2 PuN	−600.47	0.16608	1500–2200
2 SN(g)	402.20	0.02832	1500–2000
2 ScN	−623.31	0.19505	1500–1608
	−633.00	0.20108	1608–1812
	−663.97	0.21817	1812–2000
1/2 $Si_3N_4(\alpha)$	−366.86	0.16142	1500–1685
	−437.53	0.20341	1685–2151
2 Ta_2N	−507.18	0.16709	1500–1700
	−742.19	0.15778	1700–2200
2 TaN	−474.31	0.15318	1500–1700
	−466.16	0.14849	1700–2200
2 ThN	−747.72	0.16940	1500–1636
	−752.89	0.17257	1636–2000
1/2 Th_3N_4	−644.12	0.16771	1500–1636
	−647.98	0.17008	1636–2000
2 TiN	−670.30	0.18453	1500–1933
	−704.51	0.20221	1933–2200

(continued)

Table 13.1 (continued)

Compound	A	B	Temperature range [K]
2 VN	−420.60	0.15878	1500–1600
2 YN	−590.61	0.19579	1500–1752
	−599.83	0.20105	1752–1799
	−622.44	0.21363	1799–2200
2 ZrN	−720.59	0.18039	1500–2125
	−754.92	0.19674	2125–2200
2 AlS	294.44	−0.11569	1500–2000
2/3 Al_2S_3	−568.00	0.15001	1500–1800
2 AuS	278.81	−0.16201	1500–2000
2 BS	337.92	−0.17363	1500–2000
2 BaS	−1079.39	0.22866	1500–2000
2 BeS	−589.96	0.17256	1500–1556
	−611.01	0.18610	1556–1800
2 COS(g)	−412.97	−0.01893	1500–1800
2 CS(g)	322.08	−0.17466	1500–2000
CS_2	−10.29	−0.00748	1500–2000
2 CaS	−1090.20	0.20581	1500–1757
	−1392.34	0.37665	1757–2000
2 CeS	−1024.46	0.18871	1500–2000
2/3 Ce_2S_3	−919.93	0.19334	1500–2000
2 CrS	−426.18	0.12422	1500–1840
2 CuS	−274.06	0.06475	1500–2000
2 FeS	−240.68	0.07108	1500–1665
	−238.13	0.06953	1665–1809
	−262.98	0.08328	1809–2000
2 Ga_2S	−155.91	−0.05290	1500–2000
2 GeS	−32.06	−0.09247	1500–2000
2 H_2S(g)	−181.57	0.09971	1500–1800
2 InS	−303.39	0.11649	1500–1800
2 LaS	−1055.46	0.20803	1500–2000
2/3 La_2S_3	−944.27	0.18931	1500–2000
2 MgS	−1078.26	0.38527	1500–2000
2 MnS	−566.62	0.13535	1500–1517
	−592.24	0.15222	1517–1803
	−540.80	0.12371	1803–2000
MoS_2	−335.99	0.13865	1500–2000
2 Na_2S	−1171.24	0.53992	1500–2000
2 NdS	−1064.75	0.23701	1500–2000

(continued)

Table 13.1 (continued)

Compound	A	B	Temperature range [K]
2/3 Nd_2S_3	−915.92	0.20790	1500–2000
2 PS(g)	173.61	−0.02871	1500–2000
2 PbS	−238.28	0.11329	1500–1609
	−87.00	−0.08889	1609–1700
	−88.14	−0.08961	1700–2000
2 PrS	−1051.18	0.22780	1500–2000
1/2 Pr_3S_4	−917.89	0.21419	1500–2000
2 PuS	−1016.59	0.20174	1500–1800
2 SiS(g)	95.83	−0.15849	1500–1685
	−8.91	−0.09635	1685–2000
2 SnS(g)	38.84	−0.09065	1500–2000
2 SrS	−1032.32	0.18935	1500–1650
	−1311.36	0.35858	1650–2000
2 ThS	−925.69	0.18853	1500–1636
	−934.00	0.19359	1636–2000
2/3 Th_2S_3	−842.60	0.16861	1500–1636
	−846.57	0.17104	1636–2000
ThS_2	−745.35	0.16857	1500–1636
	−746.72	0.16942	1636–2000
2 TiS	−667.66	0.16709	1500–1933
	−701.57	0.18460	1933–2000
TiS(g)	490.92	−0.15819	1500–1933
	445.74	−0.13473	1933–2000
2 US	−799.57	0.19320	1500–2000
2/3 U_2S_3	−691.51	0.17770	1500–2000
US_2	−597.75	0.16983	1500–2000
2 ZnS(g)	8.91	0.06137	1500–2000
2 ZrS(g)	448.52	−0.14079	1500–2000
ZrS_2	−693.41	0.17331	1500–1823
1/3 Al_4C_3	−86.25	0.02702	1500–1800
B_4C	−79.76	0.00867	1500–2200
Be_2C	−136.11	0.02983	1500–1556
	−162.45	0.04670	1556–2200
1/2 C_2H_2(g)	110.52	−0.02593	1500–2000
CH_4(g)	−92.75	0.11164	1500–2000
1/2 CaC_2	−30.17	−0.01318	1500–1757
	−105.87	0.02992	1757–2200
Cr_4C	−101.50	−0.00335	1500–1793

(continued)

Table 13.1 (continued)

Compound	A	B	Temperature range [K]
1/2 Cr_3C_2	−43.79	−0.00866	1500–1600
1/6 $Cr_{23}C_6$	−79.22	−0.00125	1500–1823
Fe_3C	54.90	−0.03745	1500–1665
	49.07	−0.03395	1665–1809
	1.85	−0.00786	1809–2000
HfC	−231.56	0.00764	1500–2013
	−239.91	0.01180	2013–2200
1/3 Mg_2C_3	−61.52	0.05686	1500–2000
1/2 MgC_2	−19.27	0.03933	1500–2000
Mn_3C	−18.17	0.00417	1500–1517
	−55.36	0.02869	1517–1793
Nb_2C	−192.63	0.01085	1500–1800
NbC	−136.77	−0.00009	1500–1800
2/3 $PuC_{1.5}$	−42.96	0.00600	1500–2200
1/2 PuC_2	−22.42	−0.01123	1500–2200
SiC (cubic)	−72.88	0.00755	1500–1685
	−122.78	−0.03717	1685–2200
Ta_2C	−198.73	0.00007	1500–1700
	−196.50	−0.00121	1700–2200
TaC	−136.54	−0.00293	1500–1700
	−133.49	−0.00468	1700–2200
1/2 ThC_2	−63.47	−0.00322	1500–1636
	−65.31	−0.00209	1636–1688
	−62.47	−0.00377	1688–1773
	−60.25	−0.00503	1733–2028
	−69.63	−0.00040	2028–2200
TiC	−188.39	0.01466	1500–1933
	−207.28	0.02444	1933–2200
UC	−109.87	0.00196	1500–2200
1/3 U_2C_3	−73.64	0.00059	1500–2000
V_2C	−148.82	0.00421	1500–2000
VC	−105.06	0.01115	1500–2000
W_2C	8.10	−0.05035	1500–2200
WC	−38.13	0.00238	1500–2200
ZrC	−198.13	0.01028	1500–2125
	−218.72	0.01997	2125–2200

Note: Part of the temperature range for ThC_2 overlaps, but this is not a problem in practice.

13.2 Interaction Parameters (Data for Normal Steel)

The interaction parameters, e_i^j, for plain steel are summarised in Table 13.2. The data is taken from The Japan Society for the promotion of science, The 19th Committee on Steelmaking (1988) unless otherwise specified. The same data is also summarised in Hino and Ito (2010). For details of the applicable concentration ranges, states of system, etc., please refer to the original data books and/or papers.

Table 13.2 Interaction parameters, e_i^j, for plain steel

i	j	e_i^j	Temperature [K]	References
Ag	Ag	−0.04	1873	
Ag	Al	−0.08	1873	
Ag	C	0.22	1873	
Ag	Cr	−0.0097	1873	
Ag	O	−0.099	1873	
Al	Ag	−0.017	1873	
Al	Al	0.043	1873	
Al	C	0.091	1873	
Al	Ca	−0.047	1880	
Al	Cr	0.012	1873	Kishi et al. (1994)
Al	Cr	0.0096	1873	Ohta and Suito (2003)
Al	H	0.24	1873	
Al	N	0.015	1953	
Al	O	−1.98	1873	
Al	P	0.033	1873	
Al	Pb	0.0065	1823	
Al	S	0.035	1823	
Al	Si	0.056	1873	
Al	U	0.011	1873	
As	C	0.25	1873	
As	N	0.077	1873	
As	S	0.0037	1823	
Au	O	−0.14	1823	
Au	S	−0.0051	1823	
B	B	0.038	1823	
B	C	0.22	1873	
B	H	0.58	1873	
B	Mn	−0.00086	1873	
B	N	0.073	1873	
B	O	−0.21	1873	

(continued)

Table 13.2 (continued)

i	j	e_i^j	Temperature [K]	References
B	P	0.008	1673	
B	S	0.048	1823	
B	Si	0.078	1873	
Be	O	−1.3	1873	
C	Ag	0.028	1873	
C	Al	0.043	1873	
C	As	0.043	1873	
C	B	0.244	1873	
C	C	0.243	1873	
C	Ca	−0.097	1873	
C	Ce	−0.0026	1873	
C	Co	0.0075	1823	
C	Cr	−0.023	1873	
C	Cu	0.016	1833	
C	Ge	0.008	1873	
C	H	0.67	1873	
C	La	0.0066	1873	
C	Mg	0.07	1873	
C	Mn	−0.0084	1843	
C	Mo	−0.0137	1833	
C	N	0.11	1873	
C	Nb	−0.059	1833	
C	Ni	0.01	1823	
C	O	−0.32	1873	
C	P	0.051	1873	
C	Pb	0.0099	1823	
C	S	0.044	1823	
C	Sb	0.015	1873	
C	Si	0.08	1873	
C	Sn	0.022	1823	
C	Ta	−0.23	1833	
C	V	−0.03	1823	
C	W	−0.0056	1833	
Ca	Al	−0.072	1880	
Ca	Al	−0.054	1873	Köhler et al. (1985)
Ca	Bi	−0.15	1873	Song and Han (1998)
Ca	C	−0.34	1873	
Ca	Ca	−0.002	1873	

(continued)

Table 13.2 (continued)

i	j	e_i^j	Temperature [K]	References
Ca	Cr	0.014	1873	Köhler et al. (1985)
Ca	Cr	−0.18	1873	Song and Han (1998)
Ca	Cu	−0.023	1873	Song and Han (1998)
Ca	Mn	−0.0067	1873	Köhler et al. (1985)
Ca	Mn	−0.10	1873	Song and Han (1998)
Ca	Ni	−0.044	1880	
Ca	Ni	−0.043	1873	Köhler et al. (1985)
Ca	Ni	−0.049	1873	Nadif and Gatellier (1986)
Ca	Ni	−0.097	1873	Song and Han (1998)
Ca	O	−580	1873	
Ca	O	−780	1873	Itoh et al. (1997)
Ca	S	−140	1873	
Ca	Sb	−0.043	1873	Song and Han (1998)
Ca	Si	−0.096	1880	
Ca	Si	−0.095	1873	Köhler et al. (1985)
Ca	Si	−0.11	1873	Song and Han (1998)
Ca	Sn	−0.026	1873	Song and Han (1998)
Ca	Ti	−0.13	1873	Song and Han (1998)
Ca	V	−0.15	1873	Song and Han (1998)
Ce	Al	−2.67	1873	Diao et al. (1997)
Ce	C	−0.077	1873	
Ce	Ce	0.0039	1873	
Ce	H	−0.6	1873	
Ce	Mn	0.13	1873	
Ce	O	−560	1873	
Ce	S	−40	1873	
Co	C	0.02	1823	
Co	Co	0.00509	1873	
Co	Cr	−0.022	1903	
Co	H	−0.14	1873	
Co	Mn	−0.0042	1843	
Co	N	0.037	1873	
Co	O	0.018	1873	
Co	P	0.0037	1873	
Co	Pb	0.0031	1823	
Co	S	0.0011	1823	
Cr	Ag	−0.0024	1873	
Cr	C	−0.114	1873	

(continued)

Table 13.2 (continued)

i	j	e_i^j	Temperature [K]	References
Cr	Co	−0.019	1903	
Cr	Cr	−0.0003	1873	
Cr	Cu	0.016	1873	
Cr	H	−0.34	1873	
Cr	Mn	0.0039	1843	
Cr	Mo	0.0018	1873	
Cr	N	−0.182	1873	
Cr	Ni	0.0002	1873	
Cr	O	−0.16	1873	
Cr	P	−0.033	1873	
Cr	Pb	0.0083	1873	
Cr	S	−0.17	1873	
Cr	Si	−0.004	1903	
Cr	Sn	0.009	1823	
Cr	V	0.012	1873	
Cu	C	0.066	1833	
Cu	Cr	0.018	1873	
Cu	Cu	−0.02	1873	
Cu	H	−0.19	1873	
Cu	N	0.025	1879	
Cu	O	−0.065	1873	
Cu	P	−0.076	1673	
Cu	Pb	−0.0056	1823	
Cu	S	−0.021	1823	
Cu	Si	0.027	1873	
Ge	C	0.03	1873	
Ge	Ge	0.007	1873	
Ge	H	0.41	1873	
Ge	S	0.026	1823	
H	Al	0.013	1873	
H	B	0.058	1873	
H	C	0.06	1873	
H	Ce	0	1873	
H	Co	0.0018	1873	
H	Cr	−0.0024	1873	
H	Cu	0.0013	1873	
H	Ge	0.01	1873	
H	H	0	1873	
H	La	−0.027	1873	

(continued)

Table 13.2 (continued)

i	*j*	e_i^j	Temperature [K]	References
H	Mn	−0.002	1873	
H	Mo	0.0029	1873	
H	Nb	−0.0033	1873	
H	Nd	−0.038	1873	
H	Ni	−0.0019	1873	
H	O	0.05	1873	
H	P	0.015	1873	
H	Pd	0.0041	1873	
H	Rh	0.0056	1873	
H	S	0.017	1873	
H	Si	0.027	1873	
H	Sn	0.0057	1873	
H	Ta	0.0017	1873	
H	Ti	−0.019	1873	
H	V	−0.0074	1873	
H	W	0.0048	1873	
H	Zr	−0.0088	1873	
Hf	Hf	0.007	1873	
Hf	O	−3.2	1873	
Hf	S	−0.27	1873	
La	C	0.03	1873	
La	H	−4.3	1873	
La	La	−0.0078	1873	
La	Mn	0.28	1873	
La	O	−43	1953	
La	S	−79	1883	
Mg	C	0.15	1873	
Mg	O	−3	1873	
Mn	B	−0.0236	1873	
Mn	C	−0.0538	1873	
Mn	Ce	0.054	1873	
Mn	Co	−0.0036	1843	
Mn	Cr	0.0039	1843	
Mn	H	−0.34	1873	
Mn	La	0.11	1873	
Mn	Mn	0	1863	
Mn	Mo	0.0046	1843	
Mn	N	−0.091	1873	
Mn	Nb	0.0073	1843	
Mn	Ni	−0.0072	1843	

(continued)

Table 13.2 (continued)

i	j	e_i^j	Temperature [K]	References
Mn	O	−0.083	1873	
Mn	P	−0.06	1673	
Mn	Pb	−0.0029	1823	
Mn	S	−0.048	1823	
Mn	Si	−0.0327	1873	
Mn	Ta	0.0035	1843	
Mn	Ti	−0.05	1873	
Mn	V	0.0057	1843	
Mn	W	0.0071	1843	
Mo	C	−0.14	1833	
Mo	Cr	−0.0003	1873	
Mo	H	−0.13	1873	
Mo	Mn	0.0048	1843	
Mo	Mo	0.121	1823	
Mo	N	−0.1	1873	
Mo	O	0.0083	1873	
Mo	P	−0.006	1873	
Mo	Pb	0.0023	1873	
Mo	S	−0.0006	1873	
Mo	Si	8.05	1873	
N	Al	0.01	1953	
N	As	0.018	1873	
N	B	0.094	1873	
N	C	0.13	1873	
N	Co	0.012	1873	
N	Cr	−0.046	1873	
N	Cu	0.009	1879	
N	Mn	−0.02	1873	
N	Mo	−0.011	1873	
N	N	0	1873	
N	Nb	−0.068	1873	
N	Ni	0.007	1873	
N	O	−0.12	1873	
N	P	0.059	1873	
N	S	0.007	1853	
N	Sb	0.0088	1873	
N	Se	0.006	1853	
N	Si	0.048	1873	
N	Sn	0.007	1879	

(continued)

Table 13.2 (continued)

i	*j*	e_i^j	Temperature [K]	References
N	Ta	−0.049	1873	
N	Ta	−0.058	1873	
N	Te	0.07	1853	
N	Ti	−0.6	1873	
N	V	−0.123	1873	
N	V	−0.111	1873	
N	W	−0.002	1879	
N	Zr	−0.63	1873	
Nb	C	−0.486	1833	
Nb	H	−0.7	1873	
Nb	Mn	0.0093	1843	
Nb	N	−0.475	1873	
Nb	Nb	0	1873	
Nb	O	−0.72	1873	
Nb	P	−0.045	1873	
Nb	S	−0.046	1823	
Nb	Si	−0.01	1873	
Nd	H	−6	1873	
Ni	C	0.032	1823	
Ni	Ca	−0.066	1880	
Ni	Cr	−0.0003	1873	
Ni	H	−0.36	1873	
Ni	Mn	−0.008	1843	
Ni	N	0.015	1873	
Ni	Ni	0.0007	1873	
Ni	O	0.01	1873	
Ni	P	0.0018	1873	
Ni	Pb	−0.0023	1823	
Ni	S	−0.0036	1823	
Ni	Si	0.006	1873	
O	Ag	−0.011	1873	
O	Al	−1.17	1873	
O	Au	−0.007	1823	
O	B	−0.31	1873	
O	Be	−2.4	1873	
O	C	−0.421	1873	
O	Ca	−515	1873	
O	Ca	−313	1873	Itoh et al. (1997)
O	Ce	−64	1873	

(continued)

Table 13.2 (continued)

i	j	e_i^j	Temperature [K]	References
O	Co	0.008	1873	
O	Cr	−0.055	1873	
O	Cr	−0.031	1873	Itoh et al. (2000)
O	Cu	−0.013	1873	
O	H	0.73	1873	
O	Hf	−0.28	1873	
O	La	−5	1953	
O	Mg	−1.98	1873	
O	Mn	−0.021	1873	
O	Mo	0.005	1873	
O	N	−0.14	1873	
O	Nb	−0.12	1873	
O	Ni	0.006	1873	
O	O	−0.17	1873	
O	P	0.07	1873	
O	Pd	−0.009	1823	
O	Pt	0.0045	1873	
O	Rh	0.0136	1823	
O	S	−0.133	1873	
O	Sb	−0.023	1873	
O	Sc	−1.3	1873	
O	Si	−0.066	1873	
O	Sn	−0.0111	1873	
O	Ta	−0.1	1873	
O	Ti	−1.12	1873	
O	U	−0.44	1873	
O	V	−0.14	1873	
O	W	0.0085	1873	
O	Y	−0.46	1873	
O	Zr	−4	1873	
P	Al	0.037	1873	
P	B	0.015	1673	
P	C	0.126	1873	
P	Co	0.004	1873	
P	Cr	−0.018	1873	
P	Cu	−0.035	1673	
P	H	0.33	1873	
P	Mn	−0.032	1673	
P	Mo	0.001	1873	
P	N	0.13	1873	

(continued)

Table 13.2 (continued)

i	j	e_i^j	Temperature [K]	References
P	Nb	−0.012	1873	
P	Ni	0.003	1873	
P	O	0.13	1873	
P	P	0.054	1873	
P	Pb	0.011	1723	
P	S	0.034	1823	
P	Si	0.099	1873	
P	Sn	0.013	1823	
P	Ti	−0.04	1873	
P	V	−0.024	1873	
P	W	−0.023	1673	
Pb	Al	0.021	1823	
Pb	C	0.1	1823	
Pb	Co	0	1823	
Pb	Cr	0.02	1873	
Pb	Cu	−0.028	1823	
Pb	Mn	−0.023	1823	
Pb	Mo	0	1873	
Pb	Ni	−0.019	1823	
Pb	P	0.048	1723	
Pb	S	−0.32	1723	
Pb	Si	0.048	1823	
Pb	Sn	0.057	1823	
Pb	W	0	1823	
Pd	H	−0.021	1873	
Pd	O	−0.084	1823	
Pd	Pd	0.002	1873	
Pt	O	0.0063	1873	
Pt	S	0.032	1823	
Rh	H	0.13	1873	
Rh	O	0.064	1823	
S	Al	0.041	1823	
S	As	0.0041	1823	
S	Au	0.0028	1823	
S	B	0.134	1823	
S	C	0.111	1823	
S	Ca	−110	1873	
S	Ce	−9.1	1873	
S	Co	0.0026	1823	

(continued)

Table 13.2 (continued)

i	j	e_i^j	Temperature [K]	References
S	Cr	−0.0105	1823	
S	Cu	−0.0084	1823	
S	Ge	0.014	1823	
S	H	0.41	1873	
S	Hf	−0.045	1873	
S	La	−18.3	1883	
S	Mn	−0.026	1823	
S	Mo	0.0027	1873	
S	N	0.01	1853	
S	Nb	−0.013	1823	
S	Ni	0	1823	
S	O	−0.27	1873	
S	P	0.035	1823	
S	Pb	−0.046	1723	
S	Pt	0.0089	1823	
S	S	−0.046	1873	
S	Sb	0.0037	1823	
S	Si	0.075	1823	
S	Sn	−0.0044	1823	
S	Ta	−0.019	1873	
S	Ti	−0.18	1873	
S	U	−0.067	1873	
S	V	−0.019	1823	
S	W	0.011	1825	
S	Y	−0.275	1873	
S	Zr	−0.21	1873	
Sb	C	0.11	1873	
Sb	N	0.043	1873	
Sb	O	−0.2	1873	
Sb	S	0.0019	1823	
Sc	O	−3.7	1873	
Se	N	0.014	1853	
Si	Al	0.058	1873	
Si	B	0.2	1873	
Si	C	0.18	1873	
Si	Ca	−0.066	1880	
Si	Cr	−0.0003	1903	
Si	Cr	−0.021	1823–1923	Suzuki et al. (2001)
Si	Cu	0.0144	1873	
Si	H	0.64	1873	

(continued)

Table 13.2 (continued)

i	j	e_i^j	Temperature [K]	References
Si	Mn	−0.0146	1843	
Si	Mo	2.36	1873	
Si	N	0.092	1873	
Si	Nb	0	1873	
Si	Ni	0.005	1873	
Si	O	−0.119	1873	
Si	P	0.09	1873	
Si	Pb	−0.01	1823	
Si	S	0.066	1823	
Si	Si	0.103	1873	
Si	Sn	0.017	1823	
Si	Ta	0.04	1873	
Si	Ti	1.23	1873	
Si	V	0.025	1833	
Sn	C	0.18	1823	
Sn	Cr	0.015	1823	
Sn	H	0.16	1873	
Sn	N	0.027	1873	
Sn	O	−0.11	1873	
Sn	P	0.036	1823	
Sn	Pb	0.035	1823	
Sn	S	−0.028	1823	
Sn	Si	0.057	1823	
Sn	Sn	0.0098	1823	
Ta	C	−3.5	1833	
Ta	H	−0.47	1873	
Ta	Mn	0.0016	1843	
Ta	N	−0.685	1873	
Ta	O	−1.2	1873	
Ta	S	−0.13	1873	
Ta	Si	0.23	1873	
Ta	Ta	0.11	1873	
Te	N	0.6	1853	
Ti	Al	0.024	1873	Morita et al. (2005)
Ti	H	−1.1	1873	
Ti	Mn	−0.043	1873	
Ti	N	−2.06	1873	
Ti	O	−3.4	1873	
Ti	P	−0.06	1873	

(continued)

Table 13.2 (continued)

i	j	e_i^j	Temperature [K]	References
Ti	S	−0.27	1873	
Ti	Si	2.1	1873	
Ti	Si	1.43	1873	Ohta and Morita (2003)
Ti	Si	−0.0256	1873	Pak et al. (2005)
Ti	Ti	0.042	1873	
U	Al	0.059	1873	
U	O	−6.6	1873	
U	S	−0.53	1873	
U	U	0.013	1873	
V	C	−0.14	1823	
V	Cr	0.0119	1873	
V	H	−0.59	1873	
V	Mn	0.0056	1843	
V	N	−0.455	1873	
V	N	−0.4	1873	
V	O	−0.46	1873	
V	P	−0.042	1873	
V	S	−0.033	1823	
V	Si	0.042	1833	
V	V	0.0309	1873	
W	C	−0.15	1833	
W	H	0.088	1873	
W	Mn	0.0136	1843	
W	N	−0.079	1879	
W	O	0.052	1873	
W	P	−0.16	1673	
W	Pb	0.0005	1823	
W	S	0.043	1823	
Y	O	−2.6	1873	
Y	S	−0.77	1873	
Y	Y	0.03	1873	
Zr	H	−1.2	1873	
Zr	N	−4.13	1873	
Zr	O	−0.23	1873	
Zr	S	−0.61	1873	
Zr	Zr	0.032	1873	

References

Barin I, Knacke O (1973) Thermochemical properties of inorganic substances. Springer

Barin I, Knacke O, Kubaschewski O (1977) Thermochemical properties of inorganic substances: supplements. Springer, Heidelberg

Diao S, Han Q, Lin G, Chen D (1997) Equilibria of Ce–Al–O and Nd–Al–O in molten iron. Steel Res 68:469–474

Hino M, Ito K (2010) Thermodynamic data for steelmaking. Tohoku University Press, Sendai, Japan

Itoh T, Hino M, Ban-ya S (1997) Deoxidation equilibrium of calcium in liquid iron. Tetsu-to-Hagané 83:695–700

Itoh T, Nagasaka T, Hino M (2000) Equilibrium between dissolved chromium and oxygen in liquid high chromium alloyed steel saturated with pure Cr_2O_3. ISIJ Int 40:1051–1058

Kishi M, Inoue R, Suito H (1994) Thermodynamics of oxygen and nitrogen in liquid Fe–20 mass% Cr alloy equilibrated with titania-based slags. ISIJ Int 34:859–867

Köhler M, Engell H, Janke D (1985) Solubility of calcium in Fe–Ca–Xi melts. Steel Res 56:419–423

Kubaschewski O, Alcock CB (1979) Metallurgical thermochemistry, 5th edn. Pergamon press, pp 378–384

Morita K, Ohta M, Yamada A, Ito M (2005) Interaction between Ti and Si, and Ti and Al in molten steel at 1873 K. In: Proceedings of the 3rd international congress on the science and technology of steelmaking, Charlotte, North Carolina, USA, May 9–12 2005, pp 15–22

Nadif M, Gatellier C (1986) Influence d'une addition de calcium ou de magnésium sur la solubilité de l'oxygène et du soufre dans l'acier liquide. Revue de Métallurgie 83:377–394

Ohta M, Morita K (2003) Interaction between silicon and titanium in molten steel. ISIJ Int 43:256–258

Ohta H, Suito H (2003) Thermodynamics of aluminium and manganese deoxidation equilibria in Fe–Ni and Fe–Cr alloys. ISIJ Int 43:1301–1308

Pak J, Yoo J, Jeong Y, Tae S, Seo S, Kim D, Lee Y (2005) Thermodynamics of titanium and nitrogen in Fe–Si melt. ISIJ Int 45:23–29

Song B, Han Q (1998) Equilibrium of calcium vapor with liquid iron and the interaction of third elements. Metall Mater Trans B 29B:415–420

Suzuki K, Ban-ya S, Hino M (2001) Deoxidation equilibrium of chromium stainless steel with Si at the temperatures from 1823 to 1923 K. ISIJ Int 41:813–817

The Japan Society for the promotion of science, The 19th Committee on Steelmaking (1988) Steelmaking data sourcebook. Gordon and Breach Science Publisher, New York

Turkdogan ET (1980) Physical chemistry of high temperature technology. Academic Press, pp 5–26

Index

T. Matsushita and K. Mukai, *Chemical Thermodynamics in Materials Science*,
https://doi.org/10.1007/978-981-13-0405-7

V

W

Printed in the United States
By Bookmasters